658.5

D0835626

IMI *information*
Sandyford, Dublin 16
Telephone 2078

THE MANUFACTURING
ADVANTAGE

To my daughter Kathryn, with love

THE MANUFACTURING ADVANTAGE

Achieving Competitive Manufacturing Operations

Nigel Slack

MERCURY

First published in 1991
by Mercury Books
Gold Arrow Publications Limited,
862 Garratt Lane, London SW17 0NB

Set in Sabon by TecSet Limited, Wallington, Surrey
Printed and bound in Great Britain by
Mackays of Chatham plc, Chatham, Kent

British Library Cataloguing in Publication Data

Slack, Nigel
    The manufacturing advantage.
    I. Title
    658.8

ISBN 1–85251–038–2

PREFACE

I used to work for a chap, a manufacturing manager, who, when asked to arrange for some particularly difficult feat from his factory, used to reply, 'Do you want it good, or do you want it Tuesday? You can't have both.' Such was the devastating authority with which he said it that surprisingly no one ever contradicted him. I doubt if it was original, but it certainly was effective. All critics were rendered silent. How could anyone be so unreasonable as to expect the factory to be good at everything? If you want something new from the factory – higher output, wider range, faster delivery or whatever – something else must suffer. I never heard anyone tell him that they wanted it both good *and* on Tuesday. (I certainly would never have dared to suggest it even if I had thought of it.) This was a great pity because, with hindsight, he was quite wrong. You may not be able to have it good and on Tuesday straight away perhaps, but it is every manufacturing manger's responsibility to work towards the state where you can have it good, and on Tuesday, and with guaranteed delivery, and with as many knobs and whistles on it as the customer wants, and cheap too!

That is what this book is about – having it good and having it on Tuesday. In other words having the best of all manufacturing worlds, or at least aiming towards the state where you can. It looks not only at how manufacturing can make it 'good' (quality), and make it by Tuesday (speed), but also how it can make it on time (dependability), how it can change the way it is made (flexibility), and how it can make it cheap (cost). A chapter is devoted to each. There are also chapters on how each of these laudable aims are influenced by the way Manufacturing manages its process technology, the organisation of its human resources, and its external and internal supply networks. Finally there is a

v

chapter on how operations can devise improvement plans which put priorities on each of its aims.

It is not a book which addresses any national agenda, at least not directly. Nowhere will you find an assessment, either depressing or optimistic, of the West's current manufacturing predicament. No international statistics to prove, once again, that some countries are better than others at the supposedly simple business of making things. Certainly no calls to imitate solutions applied elsewhere. The underlying assumption here is that one of the most effective ways for any nation to achieve manufacturing success is for its individual companies to look to their own performance.

What this book does is to address itself directly to the individual Manufacturing company which is trying to succeed in a turbulent, competitive, and probably ill-understood, market-place. It asks some simple questions, which any company should be able to answer, about its own manufacturing operation, and at times it even offers some answers. It is aimed both at the practising manager who can see beyond simple and simplistic remedies, and at the student of manufacturing who is as interested in the rapidly accumulating weight of empirical evidence as he or she is in the more esoteric theories of manufacturing.

It is not an 'academic' book (although written by an academic) in the sense that its sole concerns are with the abstract or theoretical. But it is unashamedly academic in the sense that it tries to make ideas general enough to be useful outside the specific companies where they originated. It is, hopefully, practical in so much as it treats real problems and is not afraid to be prescriptive where it seems helpful and appropriate. It tries to blend the academic with the practical. Perhaps the book itself is trying to have it good *and* have it on Tuesday!

ACKNOWLEDGEMENTS

Ideas do not emerge fully formed. They build up over a long period during which they are shaped by many individuals and organisations. In the case of this book it is the many hundreds of managers from scores of firms who, through seminars and workshops, have often challenged, sometimes reinforced and usually reshaped my own initially primitive observations. Colleagues also, past and present, have been generous with their help and suggestions. I am particularly grateful to Richard Lamming of Bath University, Peter Race, Felix Schmid and Harvey Maylor of Brunel University, Mike Gregory and Ken Platts of Cambridge University, Bob Gibbon and Ros Miller of John Crane (UK) Ltd, Danny Samson of Melbourne University, Amrik Sohal of Monash University, Alexander Gosling of Invetech Ltd., Barrie Dale of Manchester University and Ray Snaddon of Witwaterstrand University. My colleagues and friends at Warwick University have been particularly helpful in making suggestions, reading drafts of the text, and, most of all, in forgiving my preoccupation with the project over many months. They are Stuart Chambers, Henry Correa, Alan Harrison, Bob Johnson, Christine Jones, Hans-Ulrich Mayer, Rhian Silvestro, David Twigg, and Derek Williams.

Ray Wild, of Henley – The Management College – had no direct hand in this book, but through his perceptive and good humoured encouragement over the years he has contributed more than anyone. To him, my thanks.

My wife, Angela, and daughter, Kathryn, put up with my incarceration in the study with fortitude. I am deeply grateful for their forbearance and deeply flattered that they missed me. Angela then had the unenviable task of word processing the manuscript. The magic of her technology could never fully compensate for the frequent changes, indecipherable handwriting and creative spelling. Her effort and support made it happen.

CONTENTS

1

THE DOCTRINE OF
COMPETITIVENESS

The manufacturing function in most companies represents the bulk of its assets and the majority of its people. But it is misleading to think of Manufacturing as mere bulk. It is the operation's very anatomy. Manufacturing is the bones, the sinew and the muscles of the company. A healthy Manufacturing function gives the company the strength to withstand competitive attack, it gives the endurance to maintain a steady improvement in competitive performance, and perhaps most important, it provides the operational suppleness which can respond to increasingly volatile markets and competitors.

A sickly manufacturing function, on the other hand, will handicap a business's performance no matter how sharp is its strategic sense. Many companies know the frustration of their best laid strategic ambitions rendered impotent by Manufacturing's inability to translate them into the kind of effective action which they should be able to expect. Strategy only means anything when it can be translated into operational action. It remains an abstract set of aspirations if it is devised in an operational vacuum. Competitive strategy cannot hope to be successful in the long term unless it expects Manufacturing's role in creating a strategic advantage to be both pivotal and direct. This means more than simply acknowledging the limitations of its manufacturing operation – though it will have limitations. It means that it must recognise the sheer competitive power which an effective manufacturing function can give the whole organisation.

This is not to say that a sense of strategic direction is unimportant. It is merely to stress that the competitive environment for most companies requires both strategic wit and manufacturing muscle. Sensible strategic direction is more than just important, it is a prerequisite for success. But it is not enough on its own. At the most basic level there is no better guarantee of long-term business success, nor is there a better defence against competitors, than simply making products better than anyone else. A healthy manufacturing function gives the organisation its manufacturing advantage. A sick one is worse than merely indifferent, it condemns the company to perpetual mediocrity.[1]

If a potent manufacturing function is the foundation of strategic success, then all companies should take a strategic view of their Manufacturing Operations. Thinking strategically about the operational side of the business is not the contradiction it once seemed. It is a recognition that the way in which an organisation manages its Manufacturing Operation has a significant effect on its ability to provide those things which mean success in the market place. Manufacturing is just too important to be left to jog along, secure in the comfort of its own routine. It needs strategic direction if its potential as the company's competitive engine is to be fully realised.

Box 1.1

Should anyone doubt the contribution of manufacturing to competitive success, consider the dominant companies (mainly Japanese) in industries such as motor cycles, domestic appliances, automobiles, and consumer electronics. At one time all of these industries were seen as making products which were mature, manufactured in high volume, typical 'cash cows'. Their manufacturing operations were considered capable of only marginal change. Developed markets existed with established market leaders. It was assumed that competition would be a matter of advertising, image and the occasional product development. Manufacturing had little to do except 'control' costs, 'maintain' scheduled delivery and keep quality levels 'acceptable'. The real initiative lay elsewhere, with Marketing, Strategy and Finance. Yet the companies who are now the market leaders did succeed in transforming their industries. Partly, it is true, because of their marketing skills and financial environment, but primarily because they saw the overwhelming advantage which could be gained from a sharper manufacturing regime.

COMPETITIVENESS THROUGH A MANUFACTURING ADVANTAGE

So what should be expected of the manufacturing function? What role should it play in company life? In all too many companies manufacturing is little more than an irritant, a drag on the company's competitive effort. Most of Manufacturing's time will be spent firefighting the many unexpected problems which are always threatening, and usually impairing, performance targets. Rarely if ever does manufacturing contribute to strategic decision making except to act as a constraint. Manufacturing is the reason why the company can't do what it really wants to do.

Contrast this with the role which Manufacturing could play. Here the competitive success of the whole company is a direct consequence of its manufacturing function having a superior performance to any competitors. Its products have a specification closer to the customer's needs than those made by any competitor, they are made and reach the customer 'error free', they are delivered in a lead-time faster than any competitor can match, and they always arrive at the time they are promised. Furthermore the manufacturing function has the confidence to change its stance, to adapt as the competitive environment itself changes.[2]

The gap between these two very different pictures of Manufacturing's role is defined by their achievements relative to two sets of people – customers and competitors.

Customers are the arbiters of what is important

Constructing a set of aims and objectives for Manufacturing is a matter of translating the needs (and potential needs) of customers into terms which are meaningful to it. It involves, for example, deciding whether price is more important to customers than lead-time, or product range or delivery dependability, or anything else. And if price is a more important factor, by how much? And which is the second most important factor? Further, do different products or product groups compete in broadly the same way, or are there significant differences in the relative importance of their needs? In other words how do customers value the things which Manufacturing can contribute to the operation's performance? A common response to this idea is to claim that all aspects of performance are important to customers. And so they might be, but not equally important. Some must have a greater significance to customers than others. Any marketing function which cannot give guidance over customer preference, and in terms which are

3

useful in setting manufacturing objectives, is simply not doing its job.

A particularly useful way of doing this is to distinguish between 'Order-Winning' and 'Qualifying' objectives.[3]

Order-Winning Objectives are those which directly and significantly contribute to winning business. They are regarded by customers as the key factors of competitiveness, the ones which are the most influential in their decision of how much business to give to the company. Raising performance in an order-winning objective will either result in more business or improve the chances of gaining more business.

Qualifying Objectives may not be the major competitive determinants of success, but are important in another way. They are those aspects of competitiveness where the operation's performance has to be above a particular level even to be considered by the customer. Below this critical level of performance the company probably won't even enter the frame. Above the 'qualifying' level, it will be considered, but mainly in terms of its performance in the order-winning factors. Any further improvement in qualifying factors above the qualifying level is unlikely to gain much competitive benefit.

Box 1.2

Listening to customers in order to define operations' objectives is easier for some companies than others. Big companies especially develop layers of management and divisionalised structures which make it difficult to broadcast the voice of the customer internally. That is assuming that they are not so big that they have lost the motivation to keep close to the customer anyway. IBM is an example of a very large company who realised that size and keeping close to the customer do not always go together.

In theory IBM is ideally placed to know what its customers' priorities are – it sells most types of computer product to most types of customers. One step to realising this potential is IBM's recently developed series of databases which marshal and distribute customer and product information. More significant maybe is their president's message to his employees, 'Get out of your offices. Find out what you're doing right. Find out what you're doing wrong. Do something about it.'

It is customers, then, who totally define what Manufacturing should consider important. Their needs should be Manufacturing's needs, their concerns should be Manufacturing's concerns. Yet customers' concerns are rarely static. They change with customers' own competitive priorities, and they respond to activity from competitors. What was regarded as acceptable performance before can be rendered inadequate by a competitor raising their own, and possibly the whole industry's, standards.

Competitors' performance defines your performance

If the first part of manufacturing's contribution to competitiveness is understanding customer needs and shaping its values and interests accordingly, the second part is to reach levels of performance which make it pre-eminent in the eyes of its customers. Yet although it is the customer who is to be impressed by the operation's achievements, it is not against the customers standards on which performance should be judged – it is against competitors'.

All improvement in performance is, at least potentially, worthwhile, but that marginal step which takes a company beyond the performance level of its competitors is by far the most valuable. The most significant boost to competitiveness will come when the performance of order-winning factors is raised above those of competitors. Conversely any reduction in the relative performance of qualifying factors will be particularly serious if it drops below the industry's 'qualifying' level of performance. It becomes, in effect, an 'order-losing' factor.

It may never be easy to assess competitors' performance with absolute accuracy, but most operations seem to spend a totally inadequate amount of time and effort tracking it. Knowing how your rivals are performing has obvious advantages in terms of anticipating their competitive position and in being able to learn from their successes and failures. But the longer term and ultimately greater benefit is in the way constant comparison with competitors establishes the idea of competitiveness itself within the operation. Yet how many people in the operation have responsibility for tracking competitors' manufacturing (as opposed to product) performance? Almost certainly very few when compared with those engaged in monitoring ones own internal performance. If a fraction of the staff time devoted to keeping (say) cost accounts was spent looking outwards towards competitive rivals, the operation would be far more likely to improve.

5

Box 1.3

Comparing operational performance with competitors is a key part of putting together any manufacturing improvement strategy. 'Benchmarking' in this way through measuring every aspect of the operation against the best of rival companies provides an invaluable basis for challenging operational complacency.

But the process is not without its longer term dangers. The main one is that by focusing on the performance of companies whose products closely resemble their own, an operation will fail to spot potential rivals who come from outside the existing industry, or alternatively be afraid to 'break the rules' by redefining the market. Sony Walkman is one of the most famous cases of this.[4]

As technologies become more accessible, some argue that simply comparing existing products with competitors' and making marginal improvements will at best provide short-lived gains unless all improvement takes account of the 'core competences' of the operation.[5] Core competences are the key skills and knowledge which are at the heart of the operation's ability to operate in its market. Sony's capacity to miniaturise, and 3M's competence in thin film coating, for example. The message is that as well as checking up on competitors' operational performance it is also valuable to look at their core competences – the set of skills which underpin their ability to perform.

The Manufacturing Advantage means 'Making Things Better'

Customers and competitors are both central to a competitive manufacturing operation because they define its aims succinctly: to satisfy one and be better than the other. A successful manufacturing operation relies on bringing a strong sense of both customers and competitors right on to the shop floor. Customers to act as a constant reminder to the operation of what aspects of competitiveness are important to it. Competitors to provide the measure against which its own performance should be judged. But more than this, by bringing a customer derived measure of importance together with a competitor derived measure of performance, priorities for improving the operation can be forumulated. Chapter 10 outlines the process of finding the gaps between importance

6

and performance and using them to drive strategic priorities.

Remember that the aim is to develop a manufacturing operation which can give the company an overwhelming advantage in its market place. A manufacturing based advantage which relies on the manufacturing function to provide the main ammunition in the competitive battle. In fact all the fundamental aspects of competitiveness are clearly within the mandate of the Manufacturing function. It has direct influence over such aspects of competitive performance as producing error-free products, getting goods to the customer quickly, invariably keeping delivery promises, being able to introduce innovative new products on a timely basis, providing a range of products wide enough to satisfy customer requirements, being able to change volumes or delivery dates to customer demands. And, always important, it determines the company's ability to offer products at a price which either undercuts the competition or gives a high margin, or both. In effect Manufacturing is the custodian of competitiveness for the whole organisation.

The implications of accepting this are far reaching for the organisation as whole, and especially for the role of the Manufacturing function within it. First, it means that Manufacturing should be seen as the central function in providing competitiveness. Not the dominant function but the prominent one – the competitive engine of the organisation. Second, it means that the expectations of the rest of the organisation for manufacturing should be very high, both in terms of the performance it expects and in the contribution it expects it to make to strategic debate. Third, it means aiming for a position where the company is better than its competitors at everything that is important. It quite simply 'makes things better' than the competition. And it has an unshakeable belief that 'making things better' than the competition is the only way to guarantee long-term competitive survival. That is what the manufacturing advantage is about.

'Making Things Better' means five things

It means *making things right* – not making mistakes, making products which actually are as they are supposed to be, products which are error free and always up to their design specification. Whichever way it's put, by doing this Manufacturing gives a QUALITY ADVANTAGE to the company.

It means *making things fast* – achieving an elapsed time between starting the manufacturing process and the product reaching the customers which is shorter than competitors. In doing this manufacturing gives a SPEED ADVANTAGE to the company.

It means *making things on time* – keeping the actual or implied delivery promise. This implies being able to estimate delivery dates accurately (or alternatively accept the customer's required delivery date), clearly communicate this to the customer, and then deliver on time. In doing this manufacturing gives the company a DEPENDABILITY ADVANTAGE.

It means *changing what is made* – being able to vary and adapt the operation, either because the needs of customers alter, or because of changes in the production process, or perhaps because of changes in the supply of resources. It means being able to change far enough, and being able to change fast enough. In doing this the manufacturing function gives the company a FLEXIBILITY ADVANTAGE.

Finally it means *making things cheap* – making products at a cost lower than competitors can manage. In the long term the only way to achieve this is by obtaining resources cheaper and/or converting them more efficiently than competitors. In doing this manufacturing gives a COST ADVANTAGE to the company.

These then are Manufacturing's five performance objectives: quality, speed, dependability, flexibility and cost. They are the basic building blocks of competitiveness as far as Manufacturing is concerned. By being better at them it contributes to overall competitiveness. Any manufacturing operation should be able to rank the relative importance of its performance objectives and judge its achievements in terms of each of them. Achieving a superior, or in the immediate term at least, an appropriate level of performance should be the major concern of its management.

Internal and External Performance

It is important to distinguish between the internal and external aspects of each performance objective. Any manufacturing operation is made up of a collection of smaller operations, where each department, unit or cell is an operation in its own right. Their performance can be judged using the same five performance objectives and they all contribute to the performance of the whole. In other words, the internal performance of each contributes to the external performance of the total operation – the performance which the customer sees.

External aspects of performance are relatively straightforward. For example an operation will want to develop 'speed' as a performance objective because its customers presumably value short delivery lead times. So if all parts of the operation are fast and responsive in their dealings with each other the total operation's ability to respond speedily to the customer is enhanced. See Figure 1.1.

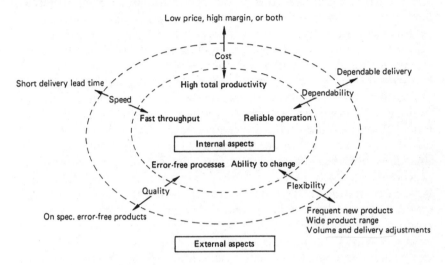

Figure 1.1: All five of manufacturing's performance objectives have their internal and external aspects

Does this mean that inside the operation speed will be the only significant performance objective? Well no, because the question then turns to the best way of improving the internal speed of the operation. There will be a number of ways. Improving quality is one. There is little point in trying to be fast if products are continually being scrapped or reworked and management are always distracted by sorting out quality problems. Similarly speed can be improved only if internal delivery dependability is high. How can you work on reducing throughput time when you can never rely on parts being processed on time? Also developing some kinds of flexibility would help. If, for example, changeover times are reduced, batch sizes can be smaller without loosing capacity. Smaller batches moving through the plant means less time spent as work in progress which, in turn, means faster throughput.

So the external benefits of (in this case) speed are that it allows the operation to offer shorter lead-times. But internally the position is more complicated. Speed is again important, but so are quality, dependability and flexibility. Internally, individual performance objectives influence, and are influenced by, each other. So

9

whereas the external aspects of performance can be taken separately, their relative merits discussed in terms of how they contribute to competitiveness, and even maybe traded-off against each other, internal performance objectives are bound together in a more intimate and complex way.

The implication of this is that it is not *what* internal performance objectives are important, but rather *why* each performance objective is important – in which way does it contribute to performance both internally and externally – and *how* can the performance of each best be improved.

Take another example, this time a company which competes primarily on price. Its manufacturing function's main external performance objective will be to reduce cost. Internally also cost issues should predominate. If all parts of the operation have high efficiencies then they all contribute to keeping total costs low. But this does not mean that cost is the only significant performance objective internally. Quality, speed, dependability and flexibility will all be important internally. Not so much to enhance external quality, speed, dependability and flexibility, but rather to reduce costs. All performance objectives are important internally. What varies is why they are important.

The next five chapters will deal with each of the performance objectives in turn and describe some of the ways in which they influence each other. Chapter 2 looks at quality, Chapter 3 speed, Chapter 4 dependability and Chapter 5 the various kinds of flexibility. Cost is left until last not because it is less important. On the contrary, it is never unimportant no matter how a company competes. Rather it is precisely because of the relationships between the internal aspects of the performance objectives. Improving any of the internal aspects of performance objectives can lead to reduced cost.

Trade-offs and improvements

Manufacturing management is sometimes portrayed as consisting almost entirely of handling trade-offs. For example, trading-off a flexible general purpose operation against a dedicated, high efficiency operation. Or trading-off high and expensive finished goods inventories against good product availability. Or expensive preventative maintenance against the reliable provision of capacity, and so on. By the law of the trade-off there is no such thing as a free lunch. Improvement in one place must be paid for elsewhere. The art is largely a matter of getting the balance right between the various objectives.

Yet if the trade-off version of manufacturing is a true representation of 'how it is', doesn't it conflict with the idea of interrelated mutually reinforcing performance objectives? Can objectives both be traded-off and reinforce each other?

Yes they can, at least partly. The trade-off argument is such a seductive one because there really is some truth in it. One way of guaranteeing fast deliveries to customers is indeed to invest in high finished goods stocks. But not the only way. Sacrificing one aspect of performance to improve another may be the most convenient solution, or it may even be the only thing one can do in the short term, but it is not the only way and it is certainly not the way to gain a long-term manufacturing advantage. It is like thinking of manufacturing performance as a seesaw where the only way to elevate one side is to lower the other. It is a view which is intrinsically limiting. Because it is possible to trade-off between objectives does not mean that improvement in one area inevitably has to be paid for elsewhere.

Think of each trade-off not as a conventional seesaw, but rather as one where the pivot as well as the beam can be moved. As with all seesaws, raising one side will indeed lower the other. And true enough one way of making an improvement in one area is by diverting resources away from, or relaxing standards in another. But here, by applying managerial effort and imagination to moving the pivot upwards, both sides of the seesaw can be raised while preserving the ability to trade-off between them. Alternatively moving the pivot could allow one side of the seesaw to be raised without lowering the other.

The pivot in this analogy is the structure, constraints, assumptions and culture of the manufacturing system itself. Raising it involves challenging long-held ideas about what really is achievable, expanding the constraints of technology, labour and systems, and persuading the whole manufacturing function that changing the 'pivot' is not only feasible but also vitally necessary. Figure 1.2 overleaf illustrates this pivot effect.

Box 1.4
There is no better illustration of this 'moving the pivot' effect than the recent transformation in how the trade-off between flexibility and productivity is viewed in batch manufacture. Some years ago the trade-off was clear. Production runs needed to be long to keep the production time which is lost on each change-over down to a minimum. The pleas for the flexibility of frequent model change-overs so that customers

could be given fast delivery were regarded as impractical. The cost would be too high. Flexibility was only to be had at the cost of more downtime and thus higher costs. The 'pivot' of this trade-off was the assumption that the time taken for a change-over was an immutable characteristic of the technology involved. The key manufacturing task was seen as working round this fixed change-over time and balancing the resulting inconvenience. The last thing anyone considered was to challenge the 'fixed' nature of the change-over itself.

Yet change-over times can be and have been reduced, dramatically in many cases. And with reduced change-over times came the possibility of increasing product mix flexibility without any reduction in productivity. (In fact, as is explained in Chapter 5, increased flexibility of this sort can actually increase productivity by making possible the adoption of just-in-time principles.) Often the engineering effort and investment required to reduce change-over times was surprisingly small. It had not been any technical difficulty preventing improvement. It had been a failure to understand the importance of tackling the constraints which stood in the way of getting 'the best of both worlds'. Raising the 'pivot' of this particular seesaw had been as much a cultural shift as anything else.

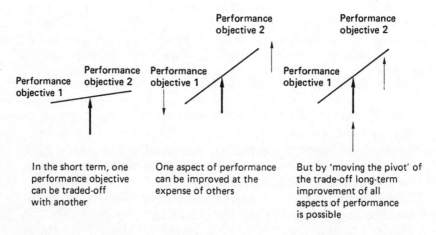

Figure 1.2: *Performance objectives*

Manufacturing focus and segmentation

The interdependence of internal performance objectives might seem to imply that a manufacturing operation can be all things to all people. It can't. The nature of the short-term trade-off is such that manufacturing is unable immediately to excel in all aspects of performance at the same time. Even if the long-term objective is to be better than all competitors at all aspects of performance, different external performance objectives require different priorities between the internal aspects of performance. This in turn leads to different types of manufacturing resource being organised in different ways. So if a company has different products or product groups competing in different ways, then its manufacturing function must take account of this in the way it is subdivided so it can maintain 'focus' on what sells the product in its market place. This segmentation of the manufacturing function into market focused sub-systems is sometimes called the 'plant within a plant' concept.

Box 1.5

The pursuit of manufacturing focus is rarely achieved without (sometimes considerable) investment and disruption to production. This is especially true when reorganisation involves several sites. Delta Crompton Cables, now one of the UK's leading cable manufacturers, was formed from the merger of two groups' cable makers, Delta and Hawker Siddeley.[6] Having decided to focus each of its plants it then had to move pieces of heavy equipment around the country in order to concentrate the manufacture of a relatively narrow range of products at each site. It was like a complex jigsaw puzzle. A machine could only move from Leeds to Derby when another machine had moved from Derby to Brimsdown. But the space at Brimsdown was not available until it had moved one of its own machines to Stalybridge. The process went on for months, but customers were always kept informed of exactly what was going on and what the company hoped to achieve from it. The policy was one of total honesty. No order was taken which they knew they would not be able to deliver.

The main savings from the reorganisation came from improvements in material usage (about 60 per cent of costs), improved simplicity of operation at each plant, and economies of scale. Runs can be longer without changing

---

13

over the machines and this reduces the chance of setting errors which can cause large material losses. Productivity of both labour and capital has also increased. Perhaps most significantly, the longer runs have allowed modern, higher speed machinery to be fully exploited. Some of its newer machines can now operate at four times the speed of older machines in the same factory.

	Product group 1	Product group 2
Products	Standard medical equipment	Electronic measuring devices
Customers	Hospitals/clinics	Medical and other OEMs
Product spec	Not high tech but periodic updates	Varies some high spec others less so
Product range	Narrow — 4 variants	Wide many types and variants some customisation
Design changes	Infrequent	Continuous process
Delivery	Customer lead-time important — ex stock	On time delivery important
Quality	Conformance/reliability	Performance/conformance
Demand variation	Financial year related but predictable	Lumpy and unpredictable
Volume/line	High	Medium to low
Margins	Low	Low to very high

External performance objectives		
Order winners	Price Product reliability	Product specification Product range
Qualifiers	Delivery lead-time Product specification Quality conformance	Delivery dependability Delivery lead-time Price

Main internal performance objectives	Cost Quality	New product flexibility Range flexibility Dependability

Figure 1.3: Different ways of competing can mean very different manufacturing performance objectives

14

The process of achieving a satisfactory manufacturing segmentation which maintains focus is often a matter of deciding which products or product groups fit together in the sense that they have similar market performance characteristics and/or they place similar demands on the manufacturing system. For example Figure 1.3 shows how two product groups manufactured by one instrument manufacturer differ in their manufacturing requirements. The first product group is a range of standard electronic medical equipment which was sold 'off the shelf' direct to hospitals and clinics. The second product group is a wider range of measuring devices which were sold to Original Equipment Manufacturers and often had to be customised to individual customer requirements. The analysis of the two product groups shown in Figure 1.3 shows that they have very different market competitive characteristics. Therefore very different external performance objectives are required from the manufacturing operation. Each product group also has different priorities for its internal performance objectives. Product group 1 needs to concentrate on cost and quality performance. All other internal performance objectives should be bent to achieving this. Product group 2 needs the flexibility to cope with a wide product range and with considerable design turbulence.

Such very different competitive needs will almost certainly require two separate focused units, each devoted to providing the things which are important in their separate markets. A point which is developed further in Chapters 7 and 8.

MANUFACTURING STRATEGY LINKS COMPETITIVENESS TO MANUFACTURING ACTIVITIES

Getting to grips with Manufacturing's performance objectives may be the essential starting point for achieving a manufacturing advantage but doesn't actually guarantee it. Such a transformation is only brought about by the operation becoming more effective at the various Manufacturing activities – the decisions and tasks which define the extent of Manufacturing's area of responsibility.

Any classification of Manufacturing's activities must be, to some extent, arbitrary. But it is useful to go back to fundamentals. These are that all manufacturing operations are built from two ingredients – technology and people. The type of technology and the type of people one chooses to have in the operation, together with their organisation and location, define the structure of the operation.

15

But manufacturing decisions do not relate solely to technology and people. The consequences of decisions taken in these two areas in effect define the way information and materials flow through the operation. Drawing departmental boundaries in a particular way or allocating particular types of technology to particular parts of the operation, for example, will shape its internal 'supply network'; that is the way requests, instructions and specifications are transmitted back through the system and materials, components and finished goods are moved forward towards customers.

It is convenient to cluster manufacturing activities around these three areas – managing process technology, the development and organisation of human resources, and the management of flow through supply networks. But admittedly it is somewhat artificial. In fact most decisions within Manufacturing will impact on all three areas. Investing in new technology for example, should certainly not be considered without its impact on skills, organisation and flow, being considered. Nevertheless most decisions will be primarily concerned with one of the three areas. That will be the criterion used here.

Technology Management defines the nature of the manufacturing technology which the company employs. Process technology decisions include the broad 'positioning' of technological capabilities within the range of available technologies, the detailed choice of particular technologies, the physical layout of the manufacturing system, and the relationship between product and process technology. Chapter 7 deals with this set of activities.

Development and Organisation includes such issues as what skills are necessary to achieve the required levels of performance, how to organise the division of these skills between groups, and the design of the manufacturing organisation. It also concerns how individuals can develop and be developed by improving the manufacturing system itself. Chapter 8 deals with this set of activities.

Managing Supply Networks at its broadest, concerns the systems which govern the collection of supply chains which connect suppliers, through the various stages of the production process to the final distribution system. This means determining the degree of vertical integration and general relationships with vendors, and the control, scheduling and storage of parts, subassemblies and finished goods. Chapter 9 deals with this set of activities.

Manufacturing Strategy

The five performance objectives – quality, speed, dependability, flexibility and cost – define what manufacturing operations are trying to achieve in order to be competitive. The various activities of Manufacturing – technology, development and organisation, and supply networks – are the ways in which resources are managed to achieve improved levels of performance. Manufacturing Strategy is the process of bringing these two sets of ideas together. It connects Manufacturing's ambitions with what it can do to bring them about. And in doing so it is making a crucial contribution to the company's success. It is making the connection between its strategic and its operational activities.

Put more formally, manufacturing strategy is the set of coordinated tasks and decisions which need to be taken in order to achieve the company's required competitive performance objectives. A company's manufacturing strategy should define its technologies, human resources, organisation, capabilities, interfaces and infrastructures. It is the last link which connects an organisation's overall business strategy to the actions of its individual resources and as such should follow directly from an understanding of competitive strategy.

FASHION, FAITH AND FAILURE

From being something of a poor relation in the business scene, Manufacturing was prodded into the management spotlight. A position it seems now rather to enjoy. It has certainly taken to examining itself with some gusto. As a result the Manufacturing function is both fashionable in itself and subject to the trends and swings of fashion. A regular deluge of ideas, techniques, theories and concepts have showered down on the Manufacturing function. Many of them come prepackaged. Total Quality Management (TQM), Just-in-Time (JIT), Optimum Production Technology (OPT), Computer Integrated Manufacturing (CIM), Total Productive Maintenance (TPM), the list seems to grow monthly.

All this attention does have some benefits. At the very least it reinforces the centrality of Manufacturing to the overall competitiveness of the company, and in doing so raises its expectations for its own manufacturing operation. But more than this, these concepts do contain more than a little good sense. The manufacturing operation has yet to be created which cannot learn something from practically any one of them. Even if they cannot

17

be applied in their entirety they can serve to stimulate, and provide fresh perspectives and fresh ideas. The problem often lies not with the concepts themselves but with the way we try to use them.

Top of the list on the negative side is the tendency of some managements to treat TQM, JIT and the like as panaceas, to swallow them whole. A danger which is not helped by the evangelical tenor adopted by the proponents of some of the concepts.[7] Yet blind faith should not be a prerequisite for the successful implementation of any concept worth its salt. On the contrary, an uncritical approach to the adoption of any of them is a sure way to sow the seeds of its eventual destruction. The problem is that, in spite of being fashionable, the manufacturing function finds itself ill prepared to cope with the attention it is attracting. Too often the hype associated with the concepts assumes more importance than the underlying messages themselves. Eventually when the concept fails to deliver all that was promised (or assumed) there is a backlash. Disillusion sets in and ideas once seen as the company's salvation are dismissed prematurely as failures. It is not the idea which has failed, it is the management of it.

The first step in avoiding the panacea pitfall is to remind ourselves, continually, that there is no universal cure-all in manufacturing – or anywhere else for that matter. We should look on JIT, TQM, *et al*, not as panaceas, but for what they are: interesting and stimulating ways of seeing operational issues, which can reshape the way we view the operation and, if we are lucky, provide the spark for creative and novel solutions to old problems. Some of the concepts may be fads, but there is a good side to fashion. It heightens our critical sense. As the fads come and go, most leave a residue of good practical sense behind them. The trick is to exploit the ideas to one's own advantage, to refine, to adapt, and to reject if appropriate.

Yet there are some ground rules on how to cope with these concepts. They come from the experiences of those companies who have made the ideas really work. What comes through is an emerging list of 'best practice' advice on what makes successful adaptation and adoption.

Avoid the hype Whether it's TQM, OPT, or JIT its merits will be subject to a certain amount of inflation and hyperbole. TQM is particularly prone to this as will be discussed in Chapter 2. The sheer motivational power of some ideas make it seductively easy to dwell on the undoubted relevance of the ideas instead of moving on to examine how one can gain solid improvement from them.

18

Put the concepts in context Don't focus on one aspect of manufacturing performance to the exclusion of everything else. Speeding up customer response, for example, may be an excellent (and fashionable) objective to pursue, but it is only one piece of the total far larger jigsaw of manufacturing performance.

Keep all improvement business related 'Flexible manufacturing', 'total quality' and so on, are not ends in themselves, they are means to the greater ends of competitive and financial success. There is no intrinsic merit in any of these things except that they can be instrumental in contributing to a clearly defined business goal.

Develop operations and people together The aim of every one of these concepts is to improve the performance of the Manufacturing system. The improvement process itself can be the most valuable of learning experiences for the operation. Yet the learning will evaporate unless the training and development of the operation's human resources is integral with the development of the manufacturing system itself. It should be the main player, not part of the supporting cast.

Finally and most important,

Don't swallow concepts whole When an operation elects to take one of these concepts on board, it must also take on the responsibility for making it work. If things go wrong it really cannot afford simply to dismiss the failure as the inadequacy of the concept – it is their problem after all. It is only sensible to make sure that any concepts taken on board are tailored to reflect the operation's individual competitive circumstances. The process of doing this is one of unbundling the concept to see what its component parts are. Only then can each be examined to see how it fits into the operation. Some parts of the concept will clearly be inappropriate, developed to fit quite different circumstances. Some will be directly applicable. Others will need adapting or modifying in some way before they can be made to work. The process is very much one of 'getting your hands dirty' inside the concept, moulding it until it fits. In the end the best operations create their own salvation, they don't rely on other people's.

PRACTICAL PRESCRIPTIONS

- Think of manufacturing as the anatomy of the organisation. It needs to be fit, lean and supple to compete.

- Define competitiveness in terms of what the operation can provide to draw a direct link between manufacturing and the market. Require the manufacturing function to provide a manufacturing advantage for the company.
- Let the relative importance of the various aspects of competitiveness for the operation be derived directly from an appreciation of customer needs. Customers are the major arbiter of what an operation should find important.
- Try to distinguish between the 'order-winning' and 'qualifying' aspects of competitiveness. Order-winners directly win business, qualifiers need to be above a certain level for the company to be considered by customers in the first place.
- If you don't know what your competitor's performance is, in operational terms, consider putting more resources into finding out.
- If you are aiming at achieving a manufacturing advantage, it means making things better than competitors. Making things better means

  Making things right – the quality advantage
  Making things fast – the speed advantage
  Making things on time – the dependability advantage
  Changing how things are made – the flexibility advantage
  Making things cheap – the cost advantage

The following are Manufacturing's performance objectives.

- Distinguish between external and internal performance objectives. The external aspects of performance objectives are seen by external customers, the internal aspects by internal 'customers'. External performance objectives can be traded-off in the short-term. Internal performance objectives are mutually supportive.
- Beware of viewing trade-offs as always losing out on some aspect of performance if another is to be improved. Work on 'moving the pivot' of the trade-off.
- If different products or product groups have different performance objectives consider segmenting the operation so as to focus on what each needs to be competitive.
- Don't swallow any of the many prepackaged manufacturing concepts whole. TQM, JIT, OPT, etc. all have messages of some relevance to all operations. Learn from them and adapt

to competitive circumstances rather than forcing them unreformed into the operation.
- Devise a manufacturing strategy which connects performance
objectives with manufacturing activities.

2

DOING IT RIGHT – THE QUALITY ADVANTAGE

Quality, to a far greater extent than any other performance objective, has an advantage which gives it great motivational power – no one disagrees with it! Who, for example, would raise their hand at a departmental meeting to declare, 'Personally I think that this emphasis on high quality work is all wrong, it really doesn't matter whether our product works or not!' On the contrary, Quality is something we all feel is worth struggling for. Quality is 'doing things right', it is 'doing what we should be doing', it is 'not making mistakes', being 'error-free'. Quality is virtuous. Of course we all believe in it!

The benefits of quality affect all aspects of performance

Yet good quality performance leads to far more than individual virtue. 'Doing things right' within the operation can transform all aspects of performance. Without errors in the manufacturing process the flow of materials through the plant can be speeded up. Or putting it the other way round: don't expect fast throughput when quality problems continually delay processing. But not only will poor quality slow down throughput, it makes it unreliable. The dependability of supply between processing stages is compromised. Not surprising that when one stage has little faith in its predecessor's ability to deliver on time, it relies on work-in-progress to provide the dependability. High work-in-progress costs money, but so does the sheer ineffective use of resources which is the inevitable result of low internal quality.

The general point is that a high level of *internal* quality performance not only makes sure that the company's products reach the customer error-free, but it also improves other aspects of internal performance, most notably speed, dependability and cost.

22

The TQM revolution

Nowhere are the pervasive and powerful arguments for increased emphasis on quality performance better summed up than in the Total Quality Management (TQM) philosophy. It is probably the most significant of the new ideas to sweep across the manufacturing scene over the last few years. That is because it makes such patently good sense. Any manager who is not convinced by the overwhelming evidence that TQM can transform an operation's performance is either remarkably isolated or irredeemably stubborn.

But it is the very power of TQM, the potency of quality as a motivating concept, which is also one of its biggest problems. It is fine to rally the organisation behind intuitively attractive (and fundamentally correct) slogans such as 'Who dares wins' or 'Right first time', but the danger is either that the whole quality effort is regarded as hype from the start or, more commonly, after initial acceptance of a quality programme, momentum becomes increasingly difficult to maintain. The effectiveness of the quality programme slows, levels off, and finally starts to fall. 'Quality droop' has set in; a gradual disenchantment with the whole process which, once started, is difficult to reverse. Better by far to avoid the pitfalls of oversell right at the introduction of the programme. (See Figure 2.1.) The trick lies in harnessing the

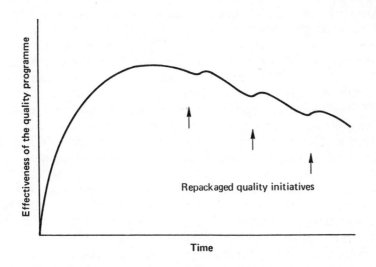

Figure 2.1: Relying on the hype of TQM will lead to 'quality droop'

tremendous motivational power which TQM can generate, without letting the 'hype' become more important than solid, steady and continuous progress towards an error-free operation.

All the warnings in Chapter 1 against relying on panaceas apply here, especially the need to understand and adapt the TQM concept to one's own circumstances. The first steps towards this are understanding the implications of the three words in TQM – Total, Quality and Management.

THE 'TOTAL' IN TOTAL QUALITY MANAGEMENT

The main difference between the traditional approach to quality and Total Quality Management is the word 'Total'. It is this which has contributed most towards the renaissance of quality. A totality of involvement which has transformed Quality Management from being at best the monitor of manufacturing mistakes to being at the centre of the drive to improve its total operations performance.

'Total' means all parts of the organisation

One of the most powerful ideas to emerge from the quality renaissance is the 'internal customer, internal supplier' concept. The idea is that the best guarantee of error free end products is for all parts of the organisation to operate in an error-free manner. Each part of the operation is an internal supplier who should provide predictable, on-specification products or service to the other parts of the organisation who are its internal customers. Errors in the service provided within the organisation will eventually affect the product which reaches the external customer. If sales order processing incorrectly records a customer order, if purchasing fails to order sufficient material, if there are errors in the design of the product, if production planning doesn't schedule sufficient capacity, then ultimately the external customer is let down just as surely as if manufacturing itself failed to make the product correctly. And the best way to ensure that external customers are satisfied is to establish the idea that every part of the organisation contributes by satisfying its own internal customers.

Each part of the operation is a link in an interconnecting network of physical and information flows – the customer for some and the supplier to others. They have the responsibility to manage these internal customer/supplier relationships by first of all defining as clearly as possible what its own and its customers' requirements are. In effect this means defining what constitutes 'error-free' service – the quality, speed, dependability and flexibility required by internal customers. The exercise replicates what should be going on for the whole operation and its external customers.

As well as helping to embed the quality imperative in every part of the operation, the internal customer concept is useful because it impacts on the 'upstream' parts of the internal supply network. These parts of the organisation, especially those who provide internal services, can originate errors which do not always become evident until later in the process. Errors show up further down the physical process route and have to be traced back to their source. Product design is a good example of this. If an error is made in the basic design of a part, that error could have been transmitted through detail design, purchasing, production engineering, and the early stages of manufacture before it becomes obvious, say at the assembly stage. A much quoted figure is that of all quality problems, about 20 per cent originate in manufacturing, 40 per cent come from vendors, and 40 per cent from design. Possibly true, but one has to start with the 20 per cent in manufacturing to find the others. It is often a matter of working backwards to find the root causes of the problems.

Box 2.1
A succinct and useful checklist to operationalise the internal customer concept comes out of the South Queensferry plant of Hewlett-Packard.[1] Its Pocket Guide has been distributed through the company.

Starting by stressing the importance of continuous improvement it goes on to suggest that each part of the organisation should ask itself seven questions fundamental to its operation.

- Who are my customers?
- What do they need?
- What is my product or service?
- What are my customers expectations and measures?
- Does my product or service meet their expectations?
- What is the process for providing my product or service?
- What action is required to improve the process?

'Total' means everyone in the organisation

Just as the quality performance of the whole company is made up of the quality performances of each part of the company, each department's quality efforts are the sum total of the individuals in it. In fact just as each department can be viewed as a process with suppliers and customers, so can every individual within the organisation. Each person influences quality, so each person has responsibility to determine their own customer-supplier requirements, and each person should see themselves as contributing to the company's overall quality performance.

All of which has implications for the style of management employed within the company generally. It is difficult to imagine the basic principles of total quality management fitting comfortably with any management style other than a relatively participative one. In the end quality performance is centred on individuals who don't make mistakes, individuals who improve the way they do the job, individuals who learn from their experience. It is not centred on procedures, techniques or philosophies. These are only the means to an end – the means to influence the individual's contribution to quality. So the whole decision-making style of the operation needs to reflect the inclusion and the contribution of its individuals.

'Total' means all costs of quality are considered

There is a cost associated with any company's quality effort, but these costs are tiny compared to the costs of not having good quality.

Traditional approaches to quality related costs were concerned mainly with trying to find the 'optimum' amount of effort to be put into improving quality. The argument being that there must be a point beyond which diminishing returns set in – the cost of improving quality gets larger than the benefits which it brings. Figure 2.2 sums this idea up. As quality effort is increased, the costs of providing the effort – through extra quality controllers, inspection procedures and so on – increases proportionally. But at the same time the cost of errors, faulty products and so on, decreases because there are fewer of them. All the extra inspectors prevent them getting out presumably.

This logic is flawed in two important respects. It underestimates one set of costs and overestimates the other. Take the cost of providing quality. The assumption is that more quality means more inspectors and so more cost. Doubling the effort put into

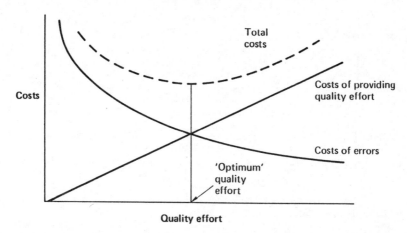

Figure 2.2: The 'optimum' quality level concept – elegant, but misleading

quality means, if not doubling the resources devoted to it, certainly a considerable increase in costs. This is not so of course. At the very heart of TQM is the idea that each of us has a responsibility for his or her own quality and is capable of 'doing it right'. This may incur some costs – training, gauges, anything which helps to prevent errors occurring in the first place – but not the steeply inclined cost curve in Figure 2.2.

The 'costs of errors' curve suffers from the opposite problem, it underestimates its true cost. This cost is usually taken to include the cost of reworking defective parts or batches, the cost of scrapping parts and materials and the loss of goodwill or even warranty costs if the defective part gets out to the customer. All these are real and important elements of the cost of poor quality, but it misses out one of the important ones: the costs associated with the disruption which errors cause. The real cost of not having quality should include all the management time wasted in organising rework and rectification. Even more important it should take into account the loss of concentration, the erosion of confidence between parts of the operation, the general disruption which quality problems cause. Do this and, even though these costs are difficult to measure, it becomes clear that error costs are higher than in Figure 2.2.

Put these two 'corrections' into the 'optimum' quality effort calculation and the picture looks very different – (Figure 2.3 overleaf). If there is an 'optimum' it is a lot further to the right, in the direction of putting more effort (but not necessarily cost) into quality.

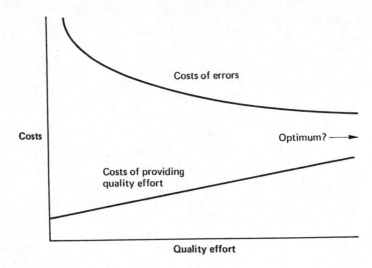

Figure 2.3: If there is an optimum, it is at a very high level of quality effort

Rather than looking for 'optimum' levels it is more revealing to see how the different aspects of quality-related costs affect each other. Think of the costs of quality under four headings.[2]

Cost of Prevention – stopping errors occurring in the first place.

- Engineering the product so that it cannot be put together incorrectly
- Checking product specifications and drawings
- Preventative maintenance of process equipment
- Developing and operating quality measurement equipment
- Administering quality procedures (for example the British Standard 5750 standards)
- Surveying quality levels, problem solving and implementing quality improvement projects
- Supplier appraisals and training
- Training and development of personnel.

Costs of Appraisal – checking to see if errors have occurred after the event.

- Product prototype testing
- Inspection and test of incoming goods
- Inspection and test of internal processes

28

- Field checks of product performance
- Processing inspection and test data

Costs of Internal Failure – coping with errors while they are still inside the organisation.

- Scrapped parts and materials
- Reworked parts and materials
- Diagnostics of quality defects and failures
- Lost production while process is stopped
- Reorganising processes and procedures after failure
- Product redesign and engineering change orders
  and finally, but possibly the most significant
- The lack of managerial concentration and focus caused by troubleshooting rather than improving the plant.

Cost of External Failure – the cost to the company of the product failing after handover to the customer.

- Warranty costs
- Servicing costs
- Product liability
- Complaints administration
  and most important in the long run, but difficult to assess
- Loss of customer goodwill affecting future business.

The useful outcome of looking at quality-related costs like this is that it helps companies to assess the relationship between the various cost categories. Of the four categories, two (costs of prevention and costs of appraisal) are open to managerial influence, while the other two (internal costs of failure and external costs of failure) show the consequences of changes in the others.

At one time it was thought that most effort should be put into appraisal so that 'bad products don't get through to the customer'. Now it is more accepted (and backed up by the experience of many companies) that preventing errors happening in the first place is a better focus for management's attention.

What seems to happen is that increased and effective effort put into defect prevention has an almost immediate positive effect on internal failure costs, followed by significant reductions both in external failure costs and, once confidence has been firmly established, in appraisal costs. Eventually even prevention costs can be stepped down in absolute terms, though prevention remains

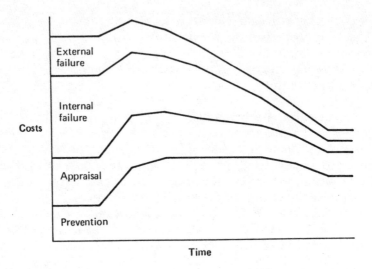

Figure 2.4: Increasing the amount of effort put into preventing errors in the process is more than offset by reducing the costs of failure and appraisal

a significant cost in relative terms. Figure 2.4 illustrates this idea. Initially total quality costs may rise as investment in some aspects of prevention – mainly training – is increased. Some reduction in total costs quickly follows, usually in months (or even weeks) rather than years.

Box 2.2
Putting realistic figures to the quality cost categories of prevention, appraisal and failure is not a straightforward matter. One study,[3] exposes a number of difficulties including the following:

- It is not easy to separate quality-related costs from those which are an integral part of the manufacturing operation anyway
- The categorisation of costs into prevention, appraisal and failure is more meaningful to Quality Managers than Operations Managers
- Costs of activities which are part-time activities of indirect staff are particularly difficult to derive
- Accounting systems are not designed to yield quality-

---

related costs and different accounting practices can distort the results in different ways

- The significance of warranty costs is difficult to gauge because they relate to earlier manufacture

Given the practical difficulties of reaching a precise breakdown of quality-related costs, it is prudent to question whether it is worthwhile at all. Certainly the evidence does appear to support the general relationship shown in Figure 2.4 and it is a useful way to think about the cost effects of quality. But that is not the same as saying that quality improvement can only be measured by including all quality-related costs in these exact categories. Better perhaps to accept the clear sense of increasing effort in the prevention category, and then take more down to earth measures of non conformance as indicators of improvement.

'Total' means all stages in quality improvement are important

Total quality management means permanently solving quality problems, and laying the foundations for further improvement in quality performance. To do this the quality improvement process must extend beyond its traditional 'monitoring and detection' role. The implications of this are for the professionals and quality 'facilitators' in the organisation – the traditional 'Quality' department. It means they need to involve themselves in the total process of defect elimination. Not only monitoring the process and recording its performance, but also analysing its performance over time, proposing solutions to any quality problems thus revealed, developing ideas for improvement, implementing the resulting changes to the process, and again monitoring the effects of the change on performance. They need to address the whole problem solving cycle (Figure 2.5 overleaf).

'Total' means all improvement is seen as a continuous process

You don't 'solve' the quality problem, you lay the foundations for further improvement. Identifying what perfect quality means to an organisation is an essential stage in understanding how far it has to improve, but perfection is something you aim for rather than ever expect to attain. It doesn't matter that one never reaches perfect quality nor should it distract from the drive for continual improvement.

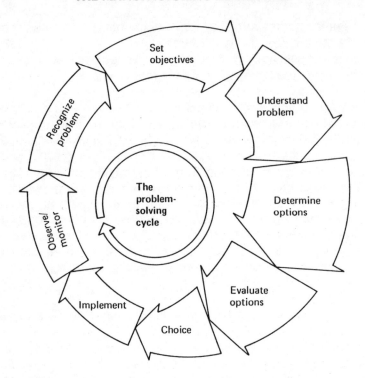

Figure 2.5:   Quality management includes all the problem solving cycle   Adapted from Cooke, S. and Slack, N., Making Management Decisions *(2nd ed.), Prentice Hall, 1991.*

Yet the race towards perfect quality is one where real competitive advantage can be gained but it does depend on an understanding of the nature of the event in which you are competing. Continuous improvement is not a sprint, it's a marathon. Success depends on maintaining an even pace rather than producing short-lived and exhausting bursts of speed.

This means that a change of attitude to improvement generally is required. Rather than success being seen as 'making big improvements' and failure 'making small improvements', success should be simply 'making an improvement'. But making an improvement every month! It is the *momentum* of improvement rather than the *rate* of improvement which is important. Momentum carries the company forward continuously, inexorably. The tortoise, after all, did eventually beat the hare!

Box 2.3

A typical experience is that of the Basingstoke UK plant of the Eaton Corporation's Truck Component Group. Although a low volume, high variety plant which is often regarded as the most difficult quality challenge, pressures from its customers (most notably Ford) convinced the company that a quality initiative would give it an important competitive advantage.

Using Statistical Process Control as the major driver of its quality programme, the company achieved considerable improvements. For example one of its flow lines which operated 24 hours a day was producing very high levels of scrap (internal failure costs) and causing customer complaints over delivery and quality (external failure costs). The company's response was an extensive improvement project on the line, which closed the line down for three days (prevention costs). It also decided to use this three day period as 'opportunity time' and use it for training the line operators (further prevention costs). At the end of the three days the operators proposed a layout which reduced standard time by nearly 10 per cent, proposed ways of reducing set-up time by over six and half hours and suggested a number of 'right first time' ideas to reduce the number of rejects.

All of these ideas reduced internal (and eventually external) failure costs. The company's rough estimate is that total quality costs which were 10.87 per cent of sales at the start of the initiative had been reduced to 4.7 per cent within four years. A 'significant saving which impacted substantially upon profitability'.

THE 'QUALITY' IN TOTAL QUALITY MANAGEMENT

Examining the word 'quality' is more than a question of semantics, though most of us do use the word 'quality' to mean different things in our everyday speech. Defining what is meant by quality is important for any organisation – if you don't know what it is, it's unlikely that you will be able to improve it! Companies need to arrive at a definition of quality which actually helps them measure, control and improve it. Look at some of the ways we use the word.

Quality can mean 'high specification'

Quality here usually is being used as a substitute for 'expensive' which often goes with a highly specified product. For example, 'the Rolls-Royce Corniche is at the quality end of the market . . .' Meaning that the car includes expensive luxurious materials, is made to a high specification and is aesthetically pleasing. But this is not what quality means in TQM. Nor does this use of the word help in establishing the importance of quality to the organisation. Yet many people find real difficulty in shedding this view of quality.

Quality can mean 'appropriate specification'

One of the more common definitions of quality is 'fitness for purpose'. In other words, does the product do what it is supposed to do? From the customer's point of view this is an ideal definition. If the product, in its operation and aesthetic properties, fulfils its purpose in an efficient and continuing manner, then the customer is unlikely to have any real complaint about the product's quality.

A definition of quality which works from the customer's perspective cannot easily be dismissed. Satisfying customers is, after all, the ultimate objective of all quality efforts. The problem with this definition is not its meaning as such, but its practical value in the operational setting. This definition of quality includes two concepts which are far more usefully treated separately. They are 'the product's specification' and whether the operation achieves 'conformance to that specification'. A product can be seen as poor quality because its design specification is poor, or its manufacture is poor, or both.

Quality can mean 'conformance to specification'

With the 'fitness for purpose' definition, two separate questions needed to be answered to judge the quality of the operation. First, 'is the performance and appearance of the product (i.e. its specification) as we want it to be?' Second, 'are we producing the product in such a way that its actual performance and appearance conforms to the specified performance and appearance?' This distinction is sometimes referred to as the difference between 'doing the right things' and 'doing things right'.

The first question is vitally important, of course. If the product's specification is not appropriate for the market then in the long run all other questions about quality become academic. However, such questions are not answered easily at the day-to-day operational level, and quality is essentially an operational issue. What can be judged is whether the operation is producing products which meet the specification. Not only is this an easier question to answer, it is fundamental to all operational processes; are they operating as they are supposed to be?

Specification can also be taken several ways

The idea of a product's 'specification' need not only relate to its physical properties. Indeed in those parts of the organisation which don't make anything physical but provide a service to the other parts, defining the specification of their 'product' can prove especially difficult for two reasons.

First, boundaries between what is acceptable and what is not are less evident and need to be defined. For example the central Manufacturing Engineering function in a large multiplant company had for many years fulfilled two roles. Their primary role was that of problem solver or technical troubleshooter. When a plant experienced difficulties which its own engineers could not overcome, it called in the central organisation. Its other role was that of 'pilot plant developer', taking new processes to the stage where they could be handed over to the plant's engineers for further development and commissioning. When one plant decided to tackle machine change-over times as part of a JIT programme it called upon central engineering for help. Unused to operating in an area somewhere between its traditional roles, it was slow to understand the plant management's requirements. Nor, to be fair, did plant management ever clearly specify their expectations. After some weeks of reducing change-over times, the central engineers were so enthusiastic to exercise their new found knowledge elsewhere, they attempted to persuade the management of a similar but traditionally more autonomous plant to adopt similar improvements. Seeing this as unasked-for interference in their operations, the plant complained to the functional director. So, paradoxically, central engineering had two dissatisfied customers ostensibly for exactly opposite reasons. But the fault was the same, failing to specify and agree the boundary of their service.

Second, for a service provider, it is not particularly helpful to separate service specification in its 'technical' sense from other aspects of operations performance such as speed, dependability and flexibility. For example, a company distributing industrial chemicals in the very early stages of its quality improvement programme struggled to define its service, using the same model as its sister company which manufactured bulk chemicals. This model defined specification strictly in terms of the properties and appearance of its products and dealt with such issues as delivery performance under a separate programme. After much debate the distribution company adopted a broad definition of its service. It did this because 'the customer does not draw fine distinctions between the various reasons why the service is bad. Whether it's because we have delivered the wrong goods, packed an incomplete order, damaged the goods, delivered late, failed to react fast enough, or been unable to package to the customer's requirements, it's all poor quality service'.

Decide what you mean by quality

So, two major decisions emerge from these various ways of viewing quality.

a) Does quality mean 'conformance to specification' or 'fitness for purpose', the latter including both the appropriateness of the specification and whether you achieve it?
b) Should you limit the quality issue to the 'technical' specification only, or consider all aspects of service performance, such as speed, dependability and flexibility?

When quality programmes are concerned with longer-term questions of whether the product offerings are still appropriate for the markets being served, and when each product is designed specifically for a customer, then 'fitness for purpose' is fine. It brings consideration of the customer's needs right to the centre of attention. If, on the other hand, quality efforts are centred on continually improving day-to-day operational performance, and especially if the shop floor staff are expected to participate in the programme, then 'conformance to specification' is usually more appropriate.

The second decision – a 'technical' or all-embracing definition of performance – depends on which part of the organisation is being considered. Direct production departments whose primary responsibility is manufacturing physical products are usually better served by clearly distinguishing between the various aspects of manufacturing performance since this allows for better diagnostics when problems do occur. So product quality, delivery speed, dependability, etc. are all treated separately. Those departments who either offer a mixture of products and services (maintenance for example) or are pure service providers (quality control itself) are often better off including all aspects of their service into the quality programme.

Figure 2.6 summarises the way quality can be defined by using these two dimensions. The simplest definition, conformance to specification of the technical aspects of specification, (the top left box of the matrix), is the most robust and most useful in the direct production parts of the plant. Quality here is concerned primarily with process performance. The top right definition is the longer term view with 'fitness for purpose' putting emphasis on the product strategy of the operation. If conformance to specification of all aspects of specification – the total service package – is

| | Definition | |
	Conformance to specification	Fitness for purpose
Technical specification of product	Quality is about process performance Simple to understand Relatively unambiguous Operationally practical Use on a regular basis for manufacturing departments	Quality is about product and process strategy Focuses on customers' view of quality Longer term orientation Use as a periodic check on appropriateness of specification
All aspects of operations service	Quality is about operations performance Need to agree service expectations with customers Some aspects of service subjective Use on a regular basis for departments who provide service	Quality is about operations strategy Needs a complete evaluation of operations objectives and performance Treated as quality only for motivational purposes

(Specification means ...)

Figure 2.6: *'Quality needs to be defined'*

taken as the definition, quality really concerns the organisation's 'operations performance'. A very broad definition, but not as all-embracing as the bottom right box where quality is being used to mean the whole operations strategy of the organisation.

THE 'MANAGEMENT' IN TOTAL QUALITY MANAGEMENT

TQM places the Quality function in possibly a difficult, certainly a challenging, and hopefully a newly influential position. Its role must change, of course. In a philosophy which stresses both that everyone makes a contribution to quality and that everyone must bear responsibility for quality improvement, quality professionals can no longer retain sole (or even the major) ownership of the information, tasks and techniques of quality. Their role must be wider – encompassing the whole quality planning and implementation task; and it must be more consultative – facilitating, guiding, coordinating and monitoring the company's quality improvement programme.

What makes for successful Quality Improvement programmes?

A number of factors appear to influence the eventual success of Quality Improvement programmes.

Top management support The importance of top management support goes far beyond the allocation of resources to the programme, it sets the priorities for the whole organisation. After all if the organisation's senior managers don't understand and show commitment to the programme, it is only understandable that the rest of the organisation will ask why they should do so.

But what does 'top management support' actually mean? It means above all that senior personnel must put a lot more effort at an operational level. It means that they must

- understand and believe in the link between 'doing things right' and the company's overall business objectives
- understand the practicalities of quality and be able to get over the principles and techniques (for example, Statistical Process Control) to the rest of the organisation
- be able to participate in the total problem solving process to eliminate non-conformance

- formulate and maintain a clear idea of what quality means for the organisation.

Lack of support can ruin a quality improvement programme. And acquiescence isn't enough. The biggest danger is the manager who pays only lip service to the concept of quality improvement but fails to participate in the programme. It takes only a hint of cynicism from senior management to undermine the whole quality effort.

The steering group   The task of the steering group is first to plan the implementation of the programme and second to make sure that even if it does not work itself out of a job, its role diminishes over time.

The first of these tasks involves planning the overall direction of the programme in terms of what it should achieve as it gathers pace. It involves deciding where to start the programme and who initially to involve. The group is also responsible for monitoring the programme and making sure that all the learning and experience which is accumulated as the programme progresses is not lost.

The second task is achieved by establishing self-supporting improvement groups.

Improvement groups   No one can really know a process quite like the people who operate it. The operators who work the line are often the ones who best know, for example, how to stop the nozzles blocking after half a shift, or who can predict that most adjustments will be needed after a product change. The engineers who devise the process routes are the ones who know, for example, that the Bill of Materials printouts are out of date, or who know to check with the foreman to see what is the real capacity of a machine. The people inside the system use the informal as well as the formal information networks. They not only have the experience of the process, they are most affected by changes in it. They are the ones therefore who should have a major role in improving the quality performance of the process.

But improvement groups cannot be formed, then left to fend for themselves. They cannot solve all their quality problems, even after training. They need support, technical, managerial and emotional. Some types of manufacturing are particularly difficult. For example if a manufacturing process is large, integrated and technically complex, many (though, it should be stressed, not all) of the technical sources of non-conformance are likely to be beyond the control of the immediate work force. The more straightforward the process technology, or labour intensive the processes, then generally the more scope an improvement group has. Yet even where technology makes self-generated improvement difficult, the group can usefully influence and guide the specialists called in to deal with the problem.

Success is recognised If quality improvement is so important, then success should be marked in some way. Recognising success formally stresses the importance of the quality improvement process as well as rewarding effort and initiative.

Many types of recognition schemes have proved effective – giving money rewards, lapel badges, certificates, quality dinners and so on. Remember though that participating in the development process itself (a part of their job which most managers take for granted) is seen quite differently by many in the organisation. For example, a company making lighting products in the UK has several manufacturing sites. At one site an improvement team had, together with some Manufacturing Systems Engineering help, developed a whole new cell layout together with in-process gauging devices and quality-charting methods. The company's management had decided to launch a similar series of quality improvement process at one of its other sites. The improvement group of seven operatives together with their supervisor, Production Manager and Quality Supervisor were asked to give a presentation at the other site. To quote one of the operatives, 'We had a rehearsal a few days before we went and we were all nervous – remember we aren't used to doing this kind of thing. But we took photographs of the cell and generally got our act together by the day. The presentation took place in the works canteen, and I think that we were all daunted by the sea of faces, but it seemed to go well. When they started asking us questions it was surprising how interested they were in our answers. No one has taken that much notice of me before! At the end they all applauded! That was the best reward of all.'

Training is the heart of quality improvement It is no coincidence that so many successful programmes have a Training Manager as one of its prime movers. TQM is above all an attitudinal change – the development task is fundamental to it. There are techniques to put over as well, of course. But the purpose of the techniques is solely to work towards the basic objective: the elimination of non-conformance.

In a way training does not 'support' the quality improvement process, training *is* the quality improvement process. The most effective people development is not separable from the development of the operation itself. Why train people away from the processes which need their attention? Why improve the quality performance of processes without capturing the lessons learned from making the improvement?

Training must be business related Return to the principal long-term threat to the effectiveness of TQM – 'quality droop'. Usually the benefits of TQM become evident to those organisations who embrace the concept. But organisations are only human, or at any rate comprised of humans! Those companies who have already passed through, and gained the benefits of, the first stages of TQM eventually enter the second stage of quality improvement. Here the problems are different.

- Improvement slows down after initial improvements as the more obvious and more easily solved problems are ironed out
- After initial success the inevitable setbacks, as the more fundamental problems are encountered, dents moral
- Enthusiasm wanes as the novelty of TQM wears off
- Other pressures take over and move management's attention away from quality.

Some remedies to these problems we have already mentioned earlier in this chapter and in Chapter 1 we stressed that the continuous nature of improvement makes training the continuing permanent upholder of quality consciousness – but the major safeguards against flagging enthusiasm are customers and competitors. The novelty of TQM may disappear, but customers and competitors won't. For most companies, even the most successful, there will *always* be someone actively trying to outperform them, or at least the threat that someone might. And it is this continuing nature of competition which can be used to drive the quality programme. Quality improvement in itself may be satisfying, but the motivation can be short-lived. Quality improvement used directly to outpace the competition on the other hand has its own permanent rationale.

So here we are, back to the fundamental message again – competitiveness is the central objective of all Operations activity. Quality is a vital element in the package of attributes which gives the company a manufacturing advantage. But also it is the concept of a sustainable competitive lead deriving directly from a company's manufacturing advantage which will sustain the quality initiative itself.

Box 2.4

When Nissan Motors (UK) established its plant in the north-east of England in the late 1980s, its indigenous rivals cannot have failed to feel at least some foreboding. Its approach and commitment to quality was not the least of the reasons. Indeed, quality was quite explicitly put forward, along with teamworking and flexibility, as part of its core philosophy. From the beginning the company's local management decided on three guiding principles.[4]

First, any programme had to be about more than quality. It needed to be integrated into the overall company activity instead of a 'bolt-on'. Its purpose was partly instrumental. It was seen as 'a means of improving individual and team development and the participation of staff in the general day-to-day running of their work areas'.

Second, it should be a natural extension of the way teams would normally operate. Paradoxically staff at the British plant chose a Japanese term, *kaizen* or 'continuous improvement' teams rather than the Japanese 'Quality Circle', to describe their team-based activity. Teams have access to '*kaizen* workshops', areas of the plant where manufacturing staff can go to make improvements. Team orientation in effect creates the environment in which quality can prosper. For example, five minutes at the beginning of each shift is spent in the team meeting. Quality problems and potential solutions are discussed, along with the results of the product audit known as VES (vehicle evaluation system). This evaluates quantitatively the quality of several vehicles from each shift. Results are analysed and immediately fed back to the relevant teams.

Third, quality would not be swamped by an external quality bureaucracy. There is a Quality Assurance department at Nissan but its main objective is to provide support and feedback to the rest of the company. Similarly the (unavoidable) steering committee operated with the minimum necessary formality and was firmly under the chairmanship of the Director of Production.

Practical Prescriptions

- Don't be dragged along by the hype of TQM. The seeds of eventual disillusionment with TQM are sown in the early days of introducing it into the company.
- Develop the idea of internal customers and internal suppliers in all parts of the operation. Encourage all parts to analyse their own operations performance in terms of quality, speed, dependability and flexibility of their service.
- The implications of TQM for management style are far reaching. Think through possible conflicts between the management style prevailing in each part of the operation and the participative style assumed in TQM principles.
- Emphasise the 'prevention' aspects of quality costs. Monitor quality, using costs if possible, but don't let it become an exercise with a momentum of its own.
- Quality staff should be problem solvers. Make sure that they see their role as extending beyond simply monitoring quality performance.
- Develop the expectation of continuous improvement carefully and link it to a comparison with competitor's performance if possible.
- Spend some time deciding what you mean by 'quality', it helps clarify improvement programmes and saves confusion.

3

DOING THINGS FAST – THE SPEED ADVANTAGE

In Manufacturing, time is more than money; time is value, it both saves cost to the operation and gives benefit to the customer. Moving requests and materials through the operation faster makes for a leaner and more productive operation. It also brings customer request and company response closer together, giving greater satisfaction to customers and less complexity for the company. Time gained is an investment in customer satisfaction and in reduced manufacturing costs.

Northern Telecom[1] for example, have come to realise that 'everything we wanted to do to improve operations had something to do with squeezing time out of our processes . . . it became clear that what we really needed to satisfy customer needs was the ability to do things faster than ever before. We needed to reorient . . . to a whole new operating strategy in which the number one priority was time. We had always asked, "How much will it cost to deliver a quality product, and how long will it take?" Now we wanted to rephrase that question to, "How quickly can we deliver a quality product, and how much will it cost?"

SPEED INVOLVES THE WHOLE THROUGHPUT CYCLE

The external customer sees the speed of an operation as the total length of time they have to wait between asking for the product and receiving it – the 'demand' time. Call this time D. But to the operation it is the whole throughput cycle which is important because that is how long the operation will have to manage the flow of materials and information.[2] Call the total throughput time P.

44

P and D times depend on the type of manufacture

For example take a typical *make-to-stock* manufacturer such as those making consumer durables. Customer's demand time, D, is the sum of the times for transmitting the order to the company's order processing system, processing the order to the warehouse or stock point, picking and packing the order, and its physical transport to the customer – the 'deliver' cycle. Behind this visible order cycle, though, lie other cycles. Reduction in the finished goods will eventually trigger the decision to manufacture a replenishment batch. This cycle – the 'make' cycle – involves scheduling work to the various stages in the manufacturing process. Physically this involves withdrawing materials and parts from input inventories and processing them through the various stages of the manufacturing route. Behind the 'make' cycle lies the 'purchase' cycle – the time for replenishment of the input stocks – involving transmitting the order to suppliers and awaiting their delivery.

So, for this type of manufacturing, the 'demand' time which the customer sees is very short compared with the total throughput cycle. (See Figure 3.1.)

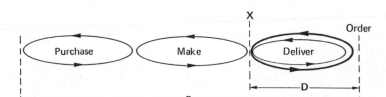

Figure 3.1:   Typical make-to-stock manufacturing throughput cycle

Contrasting with the make-to-stock company is the company which both makes and develops its products to order. Some aerospace companies for example fall into this category. Figure 3.2 illustrates the total response cycle for this type of company. Here D is the same as P. Both include the 'enquiry' cycle, a 'develop' cycle for the design of the product, followed by the 'purchase', 'make' and 'delivery' cycles.

Most companies lie between these extremes. Take for example a manufacturer of industrial couplings whose product range is far wider than its component range because it can configure components in many different ways to make its catalogue range of products. Given its wide range, the company does not hold

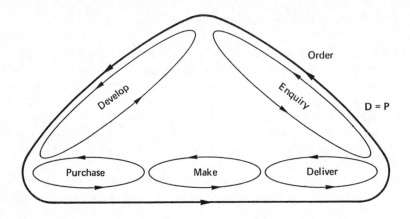

Figure 3.2:   Typical make-and-develop-to-order throughput/order cycle

finished goods inventories, instead it makes most of its components to stock and assembles its products (a relatively short process) to order. But some components are needed only infrequently because the products which they go into, although in the catalogue, have low volumes. These components are made entirely to order. In addition the company makes 'specials', couplings made to customer specifications, often needing the purchase of special components.

This company, like most, operates with more than one P and more than one D. Not that the benefits of improving P and D speed will apply only to some of its order cycles, rather that the balance of benefits will be different. It all depends on the ratio of P to D.

Box 3.1
Most manufacturing companies operate a mixture of order cycles. A UK based survey [3] found that 6.3 per cent of plants produce only for finished goods stock with supply to customers 'off the shelf'. Conversely 18 per cent make only against a firm customer order for at least part of the total order cycle. Most plants, 76 per cent, make both for stock and against customer orders.

Of the plants which make both for stock and to order, about one third are predominantly make-to-order companies with less than 20 per cent of production for stock. Only a fifth make more than 80 per cent to stock. The general impression is of a wide spread in the mix of make-to-stock versus make-to-order profiles but with a slight preponderance of make-to-order plants.

P:D ratios

In the company described above, reducing total throughput time P will have varying effects on the time the customer has to wait for demand to be filled. For many of the 'specials' P and D are virtually the same thing – the customer waits from the material being ordered through all stages in the production process. Speeding up any part of P will reduce the customer's waiting time, D. On the other hand the customer's purchasing standard 'assemble to order' products would only see reduced D time if the 'assemble' and 'deliver' parts of P were reduced and savings in time were passed on to the customer.

Generalising this in Figure 3.3, D is shown as being smaller than P, which is the case for most companies. How much smaller D is than P is important because it indicates the proportion of the operation's activities which are speculative – carried out on the expectation of eventually receiving a firm order for the work. The larger P is compared with D, the higher the proportion of speculative activity in the operation and the greater the risk the operation carries.

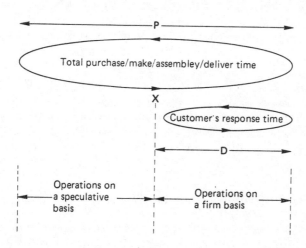

Figure 3.3: *The total throughput time has speculative and firm elements*

THE MANUFACTURING ADVANTAGE

But the speculative element in the operation is not there only because P is greater than D; it is there because P is greater than D and demand cannot be forecast perfectly. With exact or close to exact forecasts, risk would be non existent or very low no matter how much bigger P was than D. Coming at the problem another way: when P and D are equal then no matter how inaccurate the forecasts are, speculation is eliminated because everything is made to a firm order (although bad forecasting will lead to other problems). Reducing the P:D ratio becomes, in effect, a way of taking some of the risks out of manufacturing planning.

THE BENEFITS OF SPEED

The external benefits of responsiveness

Reducing the part of the total throughput cycle which lies to the right of point X in Figure 3.3 gives an obvious external benefit to the customer – they get their goods faster. In some competitive circumstances reduced delivery lead times can be vital. In others it is less important, but never totally unimportant. Indeed in many industries, such as machine tools, delivery lead time is coming more to the fore as a competitive objective. Yamazaki Mazak for example, at their machine tool plant in Worcester, have very firm views on the importance of short delivery lead times. 'Yamazaki has a broad range . . . but variety is only a trump card if it can be delivered quickly. It's short delivery times which give market share'.[4]

Paradoxically, decreasing P only to the right of point X will increase the P:D ratio, since the reduction will be a greater proportion of D than P. So the proportion of speculative activity in the company will actually be greater! This is true. The proportion of speculative activity is greater but the absolute amount is not. For an investment in the same amount of speculative production the customers get their goods sooner, competitiveness is increased, and the company gets its revenue sooner.

The internal benefits of responsiveness

Contrast an operation which moves information and materials quickly through the system with a slower one. The faster operation will have a number of advantages over its less responsive competitor.

48

Speed reduces speculative activity

Reducing throughput time prior to point X in Figure 3.3 reduces the P:D ratio and hence the proportion (and in this case the absolute amount) of speculative activity. Also as point X moves earlier in the throughput cycle (an inevitable consequence of reducing P proportionately more than D) the inventories at point X become less high value and possibly less product specific, allowing more planning flexibility.

Speed allows better forecasts

Low P:D ratios may provide some protection against poor forecasts but in addition reducing P actually makes better forecasts more likely. Events well in the future are much more difficult to forecast than imminent events. Forecast error is directly proportional to how far ahead is the event being forecast. This is especially important in intrinsically uncertain markets such as fashion garments.

Speed reduces overheads

The longer an order or a batch spends in any operation the more overheads it attracts. An order which moves quickly through the operation takes less 'looking after' than one which lingers. It needs less heating, lighting and space. It does not need as much controlling, checking and monitoring; you don't have to remember it's there! Fewer decisions need to be taken about it. It needs far less management attention.

All this means that fast throughput operations need less internal organisation and effort to support the order on its journey through the operation – less overhead, and less cost.

Speed lowers work-in-progress

When material passes quickly through the operation it cannot spend as much time in the form of work-in-progress, waiting to be processed. The time which material or information takes to move round the cycle is either taken in being processed, travelling between processing stages, or waiting to be processed. Waiting time is by far the largest element in the throughput cycle and is the obvious part to be reduced. Faster throughput means less work-in-progress and therefore less working capital.

49

Speed exposes problems

Reducing in-process inventories also has effects which have more far-reaching consequences for the operation than working capital reduction.

According to *just-in-time* philosophies of production, stocks, either of materials or information, have the effect of obscuring problems in the operation. With piles of work lying about the system it is difficult to 'see' the operation itself. Problems are hidden and improvements smothered. Take away these piles of work and the true relationship and performance differences between different parts of the operation are exposed. The very motivational structure of the operation is changed. While there is a pile of work buffering one part of the operation from the rest it is of little importance what happens elsewhere in the operation. Local difficulties will be buffered by the work-in-progress. That, at least ostensibly, is what they are supposed to be there for.

But with fast throughput and low inventories each part of the operation is exposed to the problems of the others. The efficiency of the system as a whole can be judged. Bottlenecks and weak links in the chain are exposed and can be improved. Before, inventories prevented us from even being sure where the problems were occurring. Now they are noticed immediately. But more than this, the motivational structure of the operation is sensitised. With the inventory removed it is now in everyone's interests to ensure that all parts of the operation are working well and the whole operation is motivated to improve itself rather than accepting the 'protection' which the work-in-progress gave it.

Speed can provide protection against slippage

Suppose the industry average quoted delivery lead-time is 12 weeks from order. If a company reduces its actual delivery lead-time to say, 8 weeks, it has the option of quoting the industry standard 12 weeks delivery, yet starting production as soon as it receives the firm order. If all goes well it will finish the order 4 weeks before it is needed. But if there is some problem during manufacturing, the company has 4 weeks 'buffer' which can absorb these delays. So unless the production problems are so serious that they delay throughput by more than 4 weeks, the order will be delivered on the quoted time. Thus increased internal responsiveness increases the chance that delivery will be as promised.

It's a seductive argument. In effect, buffers of inventory in the system have been traded for a buffer of time at the end of the manufacturing lead time. Two problems though: first the competitive benefits of speed are not being exploited, second, knowing that a due date *can* be missed is the best way of ensuring that it *will* be missed!

Planned freeze time

The company which has reduced D from 12 to 8 weeks might use the time it has gained in another way. Rather than start the production process as soon as it receives the order it might choose to start production 4 weeks later. Doing this it will always know exactly what it will be making 4 weeks ahead. The ability to have a 4-week 'freeze' time means that all the resources necessary for manufacture can be planned well in advance. The whole production process becomes more stable as a result.

Box 3.2

It is the manufacturers who operate in uncertain and volatile markets who have the greatest need for fast throughput. Yet often the volatility of demand makes speeding up the process particularly difficult. Take Avon Cosmetics for example. Their German subsidiary distributes almost a third of a million individual items to 100,000 sales representatives every day.[5]

Speeding up throughput times has clear advantages for Avon. It allows it to maintain or improve stock availability while reducing the amount of capital tied up in inventory, and it reduces manufacturing's sensitivity to forecast error.

Their programme to do this included three separate measures. First it put effort into improving the accuracy of its forecasts by piloting each sales campaign with a small sample of representatives three weeks ahead of the main campaign. This reduced forecast error from 33 per cent to 14 per cent. Second it rationalised the range of products manufactured at each of its three European plants to reduce complexity. Third, it mounted a supplier education initiative, eventually requiring them to deliver to just-in-time principles (or alternatively to keep high safety stocks). The results are given below.

	1982 per cent	1987 per cent
Finished good inventory	100	75
Components inventory	100	47
Raw ingredients inventory	100	48
Costs of warehousing and handling	100	75
Average run length	100	82
Customer service level	94.0	99.7

The potential savings are huge

When throughput time is the criterion by which manufacturing is judged, the results can be, quite literally, shocking. Take for example a domestic appliance manufacturer (Australian as it happens, but similar examples can be found anywhere). One particular product, a domestic heater, had a total work content of around 45 minutes. This time was available courtesy of the company's Industrial Engineers' stopwatches. Many of the operations which go to make the product will happen in parallel, but no matter, the 45 minutes represents the very maximum time necessary to manufacture one product. It is also the length of time for which value is being added to the product.

Now contrast this figure of 45 minutes with the time for the components which make up the product to move through the plant. Of course each component will be different, but just take the highest value components. Now it isn't a stopwatch which is needed, it's a calendar. The throughput 'make' time – just in the plant, not including purchasing times – averaged around 5 weeks. This means that the components are being worked on for less than 0.4 per cent of the time they are in the plant. For 99.6 per cent of the time no value is being added (though cost is). Yet this exercise examined only the higher value parts which tend to hang around the plant less than the others. Furthermore the plant was actually quite well managed with a keen and imaginative management team.

The potential for speeding things up is huge in almost any plant outside the process industries, and even there it is greater than one would think. There are, of course, many reasons, good and bad, for the vast difference between processing time and throughput time. But the exercise is a crude way of making two points. First, the sheer magnitude of potential for improvement should convince even the most reluctant management that something can be done.

Second, the many 'reasons' for slow throughput can be used not as excuses but as guides to where improvements can be made. Draw up a list of all the excuses and use it as a 'hit list' of problems to be tackled.

Box 3.3

Like other leading electronic manufacturers, Apple Computer[6] is moving towards 'customer driven manufacturing', bringing its P time down towards its D time, and in spite of increasing product variety (over 1,200 configurations) has achieved significant speed based improvement. In its Cork factory in the Republic of Ireland incoming orders take only two days to ship, a fraction of previous performance.

Improvements in the integration of the forecasting process with manufacturing have helped, but of longer term importance is the company's work in shortening response times from its suppliers. Ninety per cent of the cost of a Macintosh personal computer is in purchased components.

The company's policy is to move towards 'virtual integration' – trying to achieve all the benefits of vertical integration without the capital investment needed to build and develop components themselves. (See also Chapter 9 for more discussion of this idea.) Apple has set up a rigorous supplier certification programme which covers not only pricing and quality but also the suppliers' ability to respond quickly with technical support and fast deliveries to any of its plants worldwide.

SPEEDING UP THE OPERATION

So why is the total throughput cycle so long compared with the time actually worked on the product? Take a simplified throughput cycle such as that shown in Figure 3.1 and examine what happens both to information passing back along the supply chain and to materials moving forward.

In the 'Deliver' cycle Customers pass orders, perhaps through a salesperson who might batch orders before passing them on, perhaps by mail, perhaps by electronic means, to an order processing operation. A decision is taken when to dispatch the order, or part order if not all the goods are available. In a make-to-order operation decisions will need to be made on how the order fits into the schedule. In a make-to-stock operation the order will need picking, checking, packing, loading and physically transporting to the customer. The elapsed time for all this will probably be significantly more than the sum of the activities which make up the order cycle, for all the following reasons.

- Inherently slow data transmission media
- Batching up of work for processing
- Decision making delays over credit control, part orders, scheduling of work etc.
- Erratic work flow resulting from variable demand
- Delays to ensure full loads in transportation
- Reordering of products which fail quality standards
- Overloading in one part of the order cycle.

In the 'Make' cycle Factory orders are generated either from the order processing operation or from the finished goods warehouse. The orders are incorporated into the Manufacturing Planning and Control (MPC) system which schedules work in the individual departments. In its appointed time each batch of parts of products moves through the process until fully manufactured.

Again the actual throughput time will exceed the sum of the activities, probably by a large margin as explained previously. These are some of the reasons:

- Delays for the MPC system to be run
- A backlog of work on the factory because of demand fluctuations
- Large batch sizes because of long process set-up times
- Hold ups while priority and urgent jobs jump the queue
- Delays because of the late arrival of parts from other departments
- Bottlenecks in the flow of production
- Rework of components or products which fail quality standards
- Transportation delays in moving batches around the plant.

In the 'Purchase' cycle Materials and component parts are ordered from the company's suppliers. The cycle then more or less replicates the 'deliver' cycle. All the things which caused delay there will apply here, but this time the company is the customer. It is on the receiving end of the delays.

There are general lessons for a fast operation

Look at the reasons for delay in each part of the throughput cycle – the lessons are similar.

a) Put effort into speeding up actual process time only where major savings look possible. Instead, focus on eliminating the 90 odd per cent of the throughput time when no value is being added to the product. Do this by rethinking overall methods and procedures.

b) Simplify and streamline decision making. Do this first by identifying how many times some kind of formal decision or approval is needed during the throughput cycle. Then question whether every decision is strictly necessary. Can some be eliminated? Can some be made by exception? Finally, where decisions are necessary, direct decision making to the lowest competent authority.

c) Put things close together. Distance, whether too many metres between process stages or too many miles between the plant and suppliers, is the enemy of speed. Transportation obviously takes longer, but more than that, separation reduces the imperative for speed; out of sight is out of immediate concern.

d) Try to protect as much of the operation as possible from unexpected variation in demand and use the resulting predictability to balance capacity. A smooth flow of information or materials depends on eliminating bottlenecks.

e) Stress internal dependability. There is little point in protecting the operation from external uncertainty if internal delays are caused by machine breakdown, labour shortage or missed internal delivery times. Short lead-times mean devising realistic internal schedules and sticking to them. The next chapter deals further with the benefits of dependability.

f) Quality failures delay throughput twice over. Once because of the rework which will be needed, and again because of the general confusion and replanning which they cause. Effort put into improving quality performance reinforces efforts to improve speed performance.

g) Batching of work is one of the major causes of delay in operations. Large batches are usually a direct consequence of long changeover and set-up times. Inflexible technology is the main culprit but is also the main area of potential improvement. Chapter 5 describes how improving certain types of flexibility also improves throughput speed.

h) Get everyone thinking in terms of speed. If an operation seriously values speed as an important competitive attribute, it should measure its performance in terms of throughput times. Only then will it become important in terms of how the operation sees itself.

Be careful though of how the speed of the operation is measured. Use 'throughput efficiency' (TE) as the preferred measure. Where

$$TE = \frac{\text{processing time}}{\text{throughput time}}$$

it keeps improvements in perspective. For example, a batch of work, with a setup and processing time of 16 hours, takes 8 weeks to work its way through the plant. If after speeding up the flow through the plant, the throughput time is reduced to 5 weeks the improvement seems impressive. Yet the TE of the plant has in fact improved only from 5 per cent to 8 per cent; a move in the right direction, but still a long way to go.[7]

SPEEDING UP NEW PRODUCT 'TIME TO MARKET'

Go down to your nearest electrical goods shop to see the effect of speeding up the product development process. Are the video recorders, vacuum cleaners, TVs, the same models which were on display twelve or even six months ago? Probably not. A good proportion at least will be new. One reason for this increasingly fast rate of product innovation is down to the way technology development is speeding up. Typical is a quote from an information technology company executive:

Our product development teams can render our products obsolete far faster than we can get them to market. It is the purchasing, manufacturing, distribution and marketing functions which impose the constraints on getting new products to market rather than the process of mastering the product ideas themselves.

Technological feasibility though is only part of the picture. The major motive for rapid innovation is quite simply that a stream of continually improving new products provides a particularly effective route to competitive advantage in increasingly hard-fought markets. A company which successfully manages to have new products amongst its offerings is creating at least the possibility of catching its competitors off guard. Certainly you can't ignore or dismiss a competitor whose innovation rate beats yours. Each new product is an opportunity to move ahead of competitors. The company which makes two product innovations to its competitor's one has twice the opportunity to move ahead of them.

Speedy and well-timed product introduction can be particularly effective in tough market conditions, especially in product driven industries like the motor industry. In 1991 after years of growth, even the mighty Japanese car makers were brought up sharp by recession in their export markets and a slowdown in their home market. Nissan, Toyota, and especially Honda all suffered a decline in profits. The only companies to hold out against this trend were those like Mitsubishi and Mazda with speedily developed and timely new product offerings coming to market just at the right time.[8]

Speeding up product development

Designing and developing new products from concept through to the market place is an operation in its own right. It is a production process whose 'products' are successfully developed product ideas brought to market. Admittedly the 'products' from this operation are large, expensive and relatively infrequent compared to the routine factory operation it serves. But nevertheless, the same principles apply to speed the operation up.

Include the whole development cycle The product development cycle reaches right back to the supply of basic research and information, the appointment and/or development of appropriate human resources and acquiring the necessary facilities. This is analogous to the purchase cycle in the factory operation. And as before there is little point in speeding up the rest of the development process if getting the basic 'raw materials' of development delays the process starting. The development cycle also extends forward to the successful introduction and 'bedding down' of the product in the market. Products cannot be abandoned after the 'technical' job is done, nor be allowed to suffer delays over the last lap.

Concentrate on reducing non value added Like any 'product' new product development has 'components'. They have their own routes through the operation during which they will be processed.

But again, as in the plant, the components of the final 'product' will spend longer getting through the operation than the sum of their individual processes. Again it is more productive to tackle the non value adding activities first. Stop the design request sitting on the engineer's desk for a week before it receives attention, for example; only then put effort into shortening the design process itself. The aim is for each part of the project to flow through the development operation smoothly like products in a well balanced factory.

Simplify decision making The product development operation is largely concerned with decision making and problem solving. It is a complex network of many distinct but interrelated decisions. The potential for confusion and delay is large, the necessity for coordination and speed clear. Coordination and speed are the main motives behind the increasing use of what is variously known as simultaneous engineering, concurrent engineering or parallel development. Whatever its label the idea is to take decision cycles which were taken sequentially, and overlap them by committing the resources needed for 'downstream' decision cycles before the 'upstream' decision cycle is completed. (See Box 3.4) This idea is explained in some more detail in Chapter 7.

Box 3.4[9]

One of the complex and costly stages in the development of a new automobile is the design and manufacture of the dies which press out the car's body panels. Standards of both materials and engineering tolerances make die cutting an expensive and time consuming process.

Traditionally, the body shape is fully specified before the die design and cutting process is started. The rise in activity on the die cutting machines together with the several cuts needed and the level of precision means that the process can take up to two years. However, some manufacturers have achieved up to twice the speed of development – one year instead of two – by putting die and body designers together. The die designers know the rough configuration of the body at an early stage, enough knowledge to order the materials. They can even start rough cuts on the die, keeping within the design limits set by the body designers up to that point, and then commit to finishing the dies on final completion of the body designs.

Of course this move to overlapping development stages is

---

more than just the ingenious use of early-committed resources. It depends equally on organisational forms which
bring the two sets of designers together, and a high degree
of shared understanding and communication.

Put stages close together Or in this case people rather than
stages. This must mean some form of project based organisation
structure with personnel seconded to the development project for
all or part of its life. The close physical location of the
contributing disciplines which project teams allow is a major help
in speeding up communication and decision making. The various
forms of organisation used in development activities are discussed
in Chapter 8.

Protect the process from disturbances Product development consists to a large measure of putting sometimes vague, and perhaps
original, concepts into a defined, practical form. More than any
other type of activity it needs clarity of purpose to progress
smoothly. It also needs some protection from outside uncertainty.
This means that outside influences should be filtered so that only
vital changes are allowed through to influence the project. In fact
many 'disturbances' to the smooth flow of development are
internally generated, the consequences of not taking decisions at
the beginning of the development cycle.

Stress internal dependability The most obvious source of internal
disturbance to development projects is slippage – overrun of the
activities which make up the development cycle. The next chapter
will argue that a lack of internal dependability has a far greater
effect on the smooth running of operations than might be
imagined. This holds as true for development as for more routine
operations.

Stress quality operations In the plant errors cause rework, delay
and confusion. 'Right first time' principles are just as valid in
development operations, even if what is 'right' may only become
apparent later. Being right, even in hindsight, requires effort to be
put into getting rid of as much ambiguity from the project as early
as possible. Thorough planning and agreement over timings and
resources at the start of the cycle are the most effective ways of
achieving this.

Measure performance in terms of speed Again, if 'time to market'
is an important competitive attribute the performance of the
development function should be evaluated at least partly on it,

together with technical performance, being on-time and being on budget. But with product development, speed is best measured not in terms of throughput efficiency, as in the plant, but by comparison with competitors.

Benchmarking performance against competitors is not always easy. But in most industries some idea of competitors' development lead times can be gleaned, or even implied from the frequency of their product introductions. However it is done its purpose is twofold. First it gives the development team an external competitive performance orientation. Second it embeds the centrality of time based development firmly in the team's consciousness.

PRACTICAL PRESCRIPTIONS

- Be aware of P and D times for your different products. Rank products by their P:D ratios. The higher the ratio the more the 'speculative manufacture' risk.
- Start by reducing D if you are primarily concerned with shortening customer lead times. Reduce P if you are primarily concerned to reduce costs, and/or speculative manufacture.
- Make sure that all the benefits of fast throughput are being exploited – better forecasts, less risk, lower overheads, lower w.i.p. and faster improvement.
- Get an approximation of your 'throughput efficiency' – the ratio of actual value adding time to elapsed time for materials to go through the operation.
- Concentrate first on eliminating the non value added parts of the throughput cycle.
- Simplify decision making and direct it to the lowest competent authority.
- Minimise the distance travelled of both materials and information.
- Protect the operation from all but the most vital external disturbances.
- Stress internal dependability, quality and flexibility. They all contribute to speed.
- Make speed a central performance measure in the operation.
- Treat product development in the same way as the factory operation. Speed it up by applying the same principles.

4

DOING THINGS ON-TIME – THE DEPENDABILITY ADVANTAGE

Dependability means keeping delivery promises – honouring the delivery contract with the customer. It is the other half of delivery performance along with delivery speed. The two performance objectives are always linked in some way. For example, one could achieve high dependability merely by quoting long delivery times. The difference between the expected time and the time quoted to the customer is being used as an insurance against lack of dependability within the operation. Long delivery times can, in theory, mask poor dependability.

But that is in theory. Companies which try to absorb poor dependability inside long lead times can finish up being both slow *and* having poor dependability. There are two reasons for this. First, as mentioned in the previous chapter (and as yet another example of Parkinson's Law) delivery times tend to expand to fill the time available. Attempting to discipline the operation to achieve delivery readiness in 5 weeks when 8 are available is unambitious, fails to allow the operation to contribute to competitiveness, and is almost certainly doomed to failure. Second, again from the previous chapter, long delivery times are a result of slow internal response, high work in progress, and large amounts of non value added time. All of these are the very things which cause confusion, oversights and lack of control – the root causes of poor dependability. Good dependability is helped by fast through-put, not hindered by it.

Sometimes it is the other way round. Companies are tempted to quote difficult-to-achieve *short* lead times. This then creates major problems when the operation tries to meet the delivery dates. Speed and dependability may be the two halves of delivery performance but they are fundamentally different types of performance objective. Speed is up front in the competitive process. It is quoted – defined as part of the specification for the order. It can

61

be as much a part of the product offering as its technical specification.

Customers may make some attempt to specify dependability also – by the use of penalty clauses for late delivery for example. But it remains a 'hindsight' performance objective. One which becomes important only when a history of performance, good or bad, is established. To quote the well-worn sales saying, 'Without a fast delivery promise you don't get a chance to prove how dependable you are – you don't get the order.'

Box 4.1

One study[1] which looked at the delivery performance in manufacturing operations found a number of significant points.

a) There is relatively little attention given to the monitoring of delivery dependability. *Lesson: make delivery dependability an explicit, measured and important objective for manufacturing.*

b) In all the cases studied, management estimates of delivery dependability were higher than the company's actual dependability performance. *Lesson: dependability performance is usually worse than you think. Make sure you can define it and know what it is in your organisation.*

c) External dependability is the difference between promised delivery date and actual delivery date. But often these two pieces of information never appear on the same document. *Lesson: as a piece of management control information, delivery date is only useful when compared to promised delivery date.*

d) Small changes in mix of demand placed on an operation can very seriously affect the ability of the operation to meet its delivery promises. *Lesson: understand how changes in mix affect the operation and make sure that the operation gets sufficient notice of forecast changes in demand.*

There are different measures of dependability

In principle dependability is a straightforward concept:

Dependability = due delivery date – actual delivery date

Ideally the equation should equal zero, then it's on time. Positive means it is early and negative means it is late.

It is not quite that simple though. In practice any quoted dependability performance benefits from a little investigation. For instance, which 'due date' exactly? The date originally requested by the customer or the date quoted by the operation? If there is a difference, does the customer know about it? Also there can be a difference between the delivery date scheduled by Manufacturing and that which is promised to the customer. In which case either sales are quoting unrealistic delivery dates (which has been known) or manufacturing are slow in scheduling the order on to the plant (which also has been known). Neither are delivery dates immutable. They can be changed – sometimes by customers, more often by the operation. If the customer wants a new delivery date, should that be used to calculate delivery performance? (possibly); or if the operation has to reschedule delivery, should the changed delivery date be used? (hardly ever). It is not uncommon in some circumstances to find four or five arguable due dates for each order.

Nor is the actual delivery date without its complications. For example, delivery where? Delivery to the customer perhaps, which is all the customer is interested in. But part of that process is possibly out of the operation's control, because of third party transportation. Perhaps then dispatch time, which is easier to measure? Or the finish of processing, which Manufacturing will see as relevant? Finally, what is late? Should delivery in the same month, week, day or hour be taken as 'on time'?

Clearly the answers to all these points will depend on circumstances – industry norms, what information is available, what the measure is going to be used for and so on. But the measure should be made to fit the circumstances; often it doesn't. (See Box 4.2). Choose whatever dates actually give a measure which reflects what you are trying to assess. And use more than one measure if you can use them to diagnose where dependability problems are occurring. For example, if on average internal dependability is fine, but customers' arrival dependability is less good, the problem probably lies in the shipping and distribution area. Finally emphasise whichever measure most faithfully represents customer expectations. Because that is what dependability is really concerned with – customer expectations. They need to be managed, they need to be met, and customers need to be reminded that they have been met.

Delivery Integrity is as important as dependability

Managing customer expectations can be as important as the actual performance itself. Sometimes the first indication that a customer has not received a delivery on time is when Sales receive an irate phone call direct from the customer. Many sales people complain that given sufficient notice of a late order they can do something to help reduce the impact on the customer. A simple early warning system which alerts everyone to likely problems ahead is all that is necessary. Bowing to the inevitable late on in the delivery cycle is far more disruptive than an early admission of problems together with sensible customer management.

The idea of keeping faith with the customer can be called 'Delivery Integrity' as opposed to delivery dependability. It is clearly a related concept and could be almost as important. Good delivery integrity can go some way to compensate for poor dependability, but should be seen as an occasional recovery tactic, not 'the way we deal with unpunctuality'.

Make the standard 100% dependability

Dependability has a number of things in common with Quality. It is a 'conformance' measure. But conformance to time rather than specification. It is an attribute which influences customer satisfaction over the longer term rather than one which grabs an immediate sale. And, importantly, it is misleading to think of it only as an average. The average number of late deliveries may be only 1 per cent (an unbelievably good performance compared with most companies) but to the one customer in a hundred whose order does not arrive on time, lateness is 100 per cent. For this reason, if no other, there should not be any alternative but to strive for 100 per cent on-time deliveries.

Let customers know when it's on-time

Many companies just don't know how much they are delivering late. (See Box 4.2). But just as worrying is the obvious corollary, they don't know how much is on-time. If late deliveries are

64

something to apologise for, to be ashamed of and to work to rectify, then on-time deliveries can be good occasions to remind the customer of the excellence of the service. Customer perceptions must be managed as well as customer expectations.

Virtue in delivery dependability is not entirely its own reward. Failure to meet delivery dates is far more likely to be noticed (and remembered) by the customer than on-time delivery. And quite right too; being on-time is nothing more than the customer is paying for. Nevertheless it can help to reinforce the impression of genuinely good service in customer's consciousness if they are told it is good. Take care though. If one delivery out of a hundred arrives bearing the legend 'Another On-Time Delivery From X Co' it will only draw attention to the ninety-nine late ones. Performance needs to be good before you can boast about it.

Box 4.2
Good delivery performance may not be as good as it seems and might be achieved only at a price. Take for example the case of a UK engineering company supplying machinery to both home and export markets.[2] In spite of its own suppliers delivering 60 per cent of bought-in parts late, its own dependability record seemed if not good, at least reasonable by the standards of the industry. Its average arrears were less than one week's production.

However, closer examination showed that to achieve this performance the company was resorting to breaking down semi-finished machines and sub-assemblies which had later delivery dates, in order to satisfy more urgent orders. Not only this but the company regularly rescheduled its orders and judged itself not against the original but against the rescheduled delivery date. This self deception only served to mask poor performance and contribute to hiding the costly and disruptive practices which were only possible because of substantial overcapacity.

THE BENEFITS OF DEPENDABILITY

Again it is worth distinguishing between the external benefits of dependability (how the outside customer sees dependability), and its internal benefits (what internal customers gain), and how that benefits the whole operation.

The External Benefits are growing

Dependability is often seen as inevitably a 'qualifying' performance objective. (See Chapter 1 for an explanation of 'qualifying objectives.') It isn't. It is an attribute judged over the longer term by customers, but that isn't the same thing. The key question is, 'Can you win more business directly by being more dependable?' and increasingly the answer is 'Yes'.

Partly this is an inevitable consequence of customers becoming more sophisticated in their purchasing behaviour. Even if dependability is not as up-front at the moment of sale as other performance objectives, previous dependability performance always was a legitimate enquiry and is becoming a more common one. But even if the first order from a customer does not place as high an emphasis on dependability as other factors, subsequent ones will. That is the way competition is moving in nearly all industries. Customers stress delivery dependability because their own internal dependability requires it – just as yours does on your suppliers.

The Biggest Internal Benefit is Stability

Start by imagining a totally dependable operation: an operation where on any day everyone has turned up for work, they all came exactly on time, and they are doing their jobs precisely as they should be done. No one is making any mistakes. This results in all parts and information being passed from department to department exactly on time with absolute precision. All deliveries and arrivals at the operation arrive on time, the orders are 100 per cent complete, and all the items are themselves of perfect quality. All machines are dependable, they never break down or stop because of faulty maintenance, or for any other reason for that matter. All is predictable. All is perfectly dependable.

You may regard this as fanciful – and of course it is. No operation works so perfectly, nor will it ever. But for one moment, compare such an operation to your own. In this perfectly dependable operation how much of management's time would be saved in a typical working day? (I have tried this question now with hundreds of manufacturing managers. They insist that, on average, about nine hours out of every eight-hour shift would be saved!) All that time freed to concentrate instead on improving the workings of their operation rather than fire fighting? A high proportion of a manufacturing manager's time is spent coping with the consequences of poor dependability. But it is not only management time which an improvement in internal dependability would free up. Just as significant is the effect on the operation's concentration level. No longer are personnel constantly looking over their shoulders to make sure that what was supposed to happen actually is happening. No longer is their attention diverted constantly from their main purpose. With high dependability they stand a far better chance of keeping their eye on the ball, of achieving some lasting improvement.

Because of this, internally dependable operations will always have a head start over those which are less predictable. Dependability gives a stability to the operation on which further improvement can be based. In fact some argue that only when an operation has achieved some kind of dependability is it worth improving the other aspects of operations performance. Without dependability, improvements in speed, flexibility, quality and productivity never reach their full potential.[3]

From this stability come other benefits.

Less inventory Part of the reason for the build-up of inventory between stages in an operation is that it buffers each stage from the output variation of its neighbours. And indeed in-process inventory is often justified on this basis. Machine breakdown, quality reworks, lack of control, all cause delay, and what more natural reaction to a history of, say, half-day delays than to allow an input stock of half a day's production to build up. With increased dependability however there would not be any need for the 'insurance' of buffer inventory.

Faster throughput The consequence of increased dependability leading to lower inventory is that goods spend less time waiting between stages. This, as we discussed in the previous chapter, reduces throughput time for the operation with all the benefits which derive from that.

IMPROVING DELIVERY DEPENDABILITY

The relationship between lead time and dependability applies inside the operation as well. Deliberately keeping some time or capacity back can be an effective strategy for increasing dependability provided the spare time is used effectively.

For example, the assembly department of a Singaporean company manufacturing computer memory devices became increasingly frustrated at the variability of its output rates, shift by shift. Reasonably, to their way of thinking, in an eight-hour shift they should schedule eight hours of work. This, in theory, would keep all resources occupied through the shift. However, they rarely produced eight hours' worth of work. Sometimes it would be as low as seven hours' worth, sometimes almost eight hours. It varied, but on average they produced about seven and a half hours' worth of work in the eight-hour shift. Machine breakdown, parts shortage and the general day-to-day problems of the department always seemed to prevent them getting a full eight hours' worth of work out of a shift.

Output was not their major problem; in fact the factory had slight surplus capacity. More serious was their inability to predict exactly how much they would make day by day. Poor internal dependability was wreaking havoc with schedules. Most weeks started by having to reschedule the whole plant because the department had failed to produce all the work scheduled to it. In spite of its surplus capacity the plant was barely meeting schedule, and cost variances were poor by company standards. Corporate disapproval prompted radical measures.

Exhibiting both insight and some courage, the company put dependability above short-term output and utilisation. It started scheduling seven hours' worth of work to each shift instead of eight. This meant that, on average, they were getting less work out of the department than before, *but at least it was dependable*. They could rely on getting their seven hours' worth of work. More importantly, they didn't waste any surplus time. At the end of each shift, they had between a few minutes and an hour of spare time after the completion of their seven hours' worth of work. They used this time wisely, by forming improvement groups on each line within the department. The task of these groups was primarily to investigate ways of increasing dependability within their section. Their suggestions were concerned with such issues as the reliability of the surface mount machines, reducing and standardising changeover times, providing alternative process routes when (more unusually as time went by) breakdown did occur, and communication with (mainly internal) suppliers.

Eventually, the various improvements suggested by these groups began to take effect. Output from the department didn't rise of course, for they were only scheduled for seven hours' worth of work each day. However, the average time left at the end of each shift gradually began to increase. Eventually it became possible to schedule seven and a quarter hours' worth of work on a shift and still be sure that it could produce it dependably.

Further improvement in dependability allowed an increase to seven and a half hours' worth of work. The department was now producing the same amount of work every day as it did on average when it was scheduled for its full eight hours. The important point, though, was that it was producing it dependably. Uncertainty had been taken out of the whole planning and control process. Eventually the department was being scheduled at close to eight hours' worth of work per day, a figure it dependably produced day after day.

The general message, then, is not to over-strain capacity but use any slack time wisely for the further improvement of dependability.

Dependable operations require dependable technology

Machines are too important a resource in manufacturing operations to be treated as they often are. Typically, both the level of care they are given and the expectations we have of them are too low. Yet the former is a direct consequence of the latter.

Why should we have such low expectations of process technology? We really are remarkably tolerant of less than perfect machine dependability. Talk with some engineers and it becomes clear that they see (and treat) machines as amiable eccentrics, amusing or irritating depending on when they choose to break down. A kick here, a quick adjustment of some part there, and it seems to work OK for a couple of hours. The satisfaction of this lies in repairing it (however temporarily) rather than stopping it ever breaking down in the first place.

Yet we have far more exacting expectations of technology when our lives depend on it. Failures in aircraft or heart–lung machines for example are treated much more seriously. And for obvious reasons – the consequences of failure are tragic. But so are technology failures in the more mundane world of manufacturing operations. Tragic in their own way: they destroy the smooth running and the predictability of the operation.

Breakdowns are far more costly, in real terms, than many operations realise. Traditionally an hour of downtime is seen as costing an hour of lost capacity, or even less if the failure is not at a bottleneck. But calibrated against the totally dependable operation described above, the cost is almost certainly far higher. It needs to include the immediate costs of replanning and generally accommodating the disruption. More important, it needs to include the longer-term corrosion of the various parts of the operation's trust in each others dependability. It is this which retards the operation's ability to improve itself.

Recognising the real cost of poor dependability has a dramatic effect on how preventative maintenance is viewed. Figure 4.1 shows the traditional approach to choosing the amount of effort to put into work which helps to stop breakdowns happening. The more effort – by employing more maintenance people, for example – the more the cost of providing that effort. But the more preventative effort is put in, the less chance of a breakdown, so the lower are the downtime costs. The sum of both these 'costs' is the total breakdown cost, which minimises at a particular level of maintenance effort – the optimum amount of maintenance.

This is a similar argument (and a similar curve) to that used to define an 'optimum' quality effort in Chapter 2. The argument is just as fallacious as it was for quality, and for the same reasons. One line on the graph is too high, the other too low.

The cost of providing maintenance care must usually include the efforts of specialist staff – but not exclusively. Many routine maintenance operations can be quite readily performed by operating staff. In fact there are distinct advantages to be gained from including the care of machines and equipment within an operator's expanded job definition, again for the same reasons as with quality. Providing preventative maintenance care need not be as costly as the line in Figure 4.1 implies.

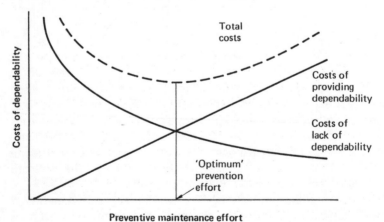

Figure 4.1: *Like the quality trade-off, the dependability-cost trade-off theory is elegant, but misleading*

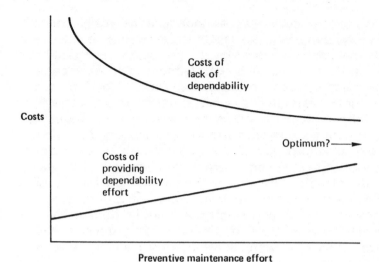

Figure 4.2: *If there is an optimum, it is at a very high level of maintenance effort*

In contrast the 'cost of breakdown' curve is too low. The real costs, as discussed previously, are far higher than assumed. The net affect is, as with the quality curve in Chapter 2, to push any 'optimum' there might be, well to the right as in Figure 4.2. In other words investment in dependability through preventative maintenance can be justified in sound economic terms.

Box 4.3

There are few better illustrations of how the cost of providing dependability can be reduced when it is tackled head on. Take Colgate Palmolive's household cleaning products plant, for example.[4] Although it had never regarded itself as a particularly maintenance orientated company it became aware of the high cost of the spares and materials holdings it was keeping in case of plant breakdown. These costs can account for up to 50 per cent of total maintenance costs. A routine stock-take during an audit exercise had revealed a £33,000 discrepancy between what they thought was in the engineering stores and what actually was there. This brought into question both the accuracy of their maintenance stock levels and why they needed such a high level in the first place (over £250,000).

The plant's reaction was to introduce its maintenance control programme, whose initial objective was to reduce the value of maintenance stocks by 15 per cent. The programme involved the identification of the least dependable machines and a data collection exercise to identify their main problem areas. It also involved purchasing a relatively basic control system, which in turn meant a considerable data entry exercise. Because of this the system took more than six months to go live, but the benefits quickly exceeded expectations. Stock savings were more than twice what was expected but, even more significantly, the availability of purchasing and stock information allowed for tighter control of the whole process, and established more orderly relationships with suppliers. Because Colgate was a more reliable customer, suppliers were more cooperative – in holding stock for example.

Internally also, the benefits soon became obvious. Before, the unreliability of the maintenance stores had resulted in the production areas often holding their own safety stores. As well as duplicating stocks it also meant that identical parts had sometimes been purchased from different suppliers. With more effective control the production department ceased to hold their own stocks, obsolete parts were scrapped and eventually the stores moved into an area half the size of their old area.

All the Operation Contributes to Total Dependability

It is more than dependable technology which makes for a dependable operation. It is the combined dependability of every part of the operation. A delay in one link in the operation's internal supply chain will either have to be made up in the later stages of the process or, alternatively, knock on to delay delivery to the customer. Both are bad. The latter for obvious reasons, the former because it will mean rescheduling work downstream, lowering the dependability of their supply.

Yet there are a number of practical steps each part of the operation can take to improve dependability. They can

- Plan ahead
- Control loading
- Enhance their flexibility

72

- Monitor progress
- Develop internal suppliers.

Planning ahead prevents surprises Enquire into why something was late and nine times out of ten some occurrence which was unexpected will be blamed. When the unexpected is outside the operation's control it is unfortunate. When it is generated internally there is less excuse for not trying to avoid it. Often the 'unexpected' could have been predicted with some internal forecasting – looking forward for indications of possible trouble. The mechanisms used to do this may be simple checklists or sophisticated computer simulations. The principle is the same: look for trouble before it takes you by surprise.

Don't overload capacity Loading an operation above its operational capacity seems to be one of the surest recipes for missing internal delivery dates. Excess load gives ideal conditions for lack of control, overlooked due dates and confusion generally. Disciplined order input control is vital to an ordered progression of work through the operation.

Flexibility can localise disruptions Certain types of flexibility (see the next chapter for a full description of the various types of manufacturing flexibility) can serve to localise any disruptions when they do occur, by providing alternative processing capability. For example, if a machine does break down (in spite of preventative maintenance), provided there is sufficient flexibility in other parts of the department then the job can be rerouted and completed in time for it to be transferred on time. The types of flexibility particularly useful here will be a sufficiently wide range of processing capabilities on the part of both machines and people, the ability to change over the machine quickly, and the ability to provide the extra capacity needed on those parts of the operation which take on the extra load.

Be clear about using flexibility in this way though. It doesn't prevent disruption; the department where the breakdown, or whatever, happens will still be thrown out of its planned routine. It would certainly be better if the disruption never happened. But if it does, flexibility can limit the damage.

Monitor progress closely What you don't know might not hurt you immediately but inevitably the mess will be worse to sort out the longer it is left. Yet surprisingly a common cause of lateness seems to be overlooked internal delivery dates. Every day, for which internal lateness is not picked up is a day less to do something about it. A workable monitoring system is essential and tends to be self-reinforcing. As internal dependability increases and flow becomes more predictable, it is easier for internal customers to signal late deliveries.

Emphasise internal supplier development   Who has the most to gain from a department or cell's dependable delivery? Its internal customers. So who should take a prominent role in helping to develop the department's capability to deliver on time? Again its internal customers. It is a microcosm of the supplier development imperative at the company level. (See Chapter 9.)

Initially internal customers can help by monitoring the delivery performance of their suppliers, making clear the importance of on-time delivery, and indicating relative priorities – pointing out for example the jobs which really cannot be late. Later it can be a matter of fine tuning – improving communications, holding joint improvement team meetings and so on. But always the underlying attitudinal shift is maintained: lateness is more than just an internal performance measure, it's letting the internal customer down.

Improve dependability, then speed

It should be clear that speed and dependability affect each other both internally and externally. The linkage between the two performance objectives becomes important when a choice is being made about which to improve first. This is best illustrated by examining the difference between quoted average delivery lead times and actual average delivery lead times. See Figure 4.3.

The delivery performance of many organisations is poor. In other words the average actual lead time achieved is longer than the quoted lead time. If both speed and dependability need to be improved, it is little use shortening lead times without increasing dependability. The first step therefore must be to move average actual lead times towards quoted lead times. In fact, since there is likely to be some spread around the average lead time, it is necessary to move it beyond the quoted lead time so that all or most deliveries take place within the promised time. Only then should quoted lead times be shortened, but not by so much as to reduce the external dependability of the operation.

So, improvement can progress, though always with dependability preceding speed. Ideally, as the improvement process progresses, the spread of lead times will reduce. That is, the whole process will get more predictable. Thus the steps of the improve-

74

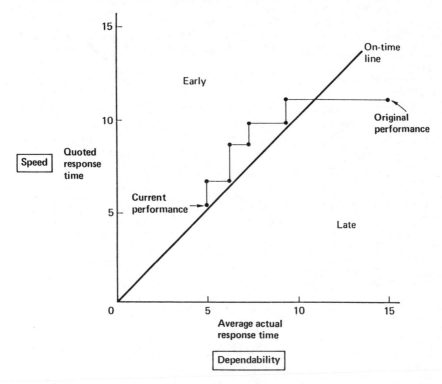

Figure 4.3:   Get dependability under control before trying to speed up delivery

ment path will get smaller, though not necessarily more frequent since by this time improvement will become increasingly difficult.

Now a new decision becomes evident, should all the improvement in actual lead-time be 'given' to the customers. One strategy is to immediately transfer all gain in internal lead time reduction to the customers, in effect following the 45° line in Figure 4.4 overleaf. Conversely, at the extreme, the opposite strategy would be to keep delivery quoted times the same and absorb all benefits within the manufacturing system. Most likely a mid course, but one closer to the 45° line, will be appropriate. Figure 4.4 illustrates these three strategies.

PRACTICAL PRESCRIPTIONS

- Delivery speed and delivery dependability are closely linked. Don't let long lead times mask poor dependability. It's the worst of both worlds.

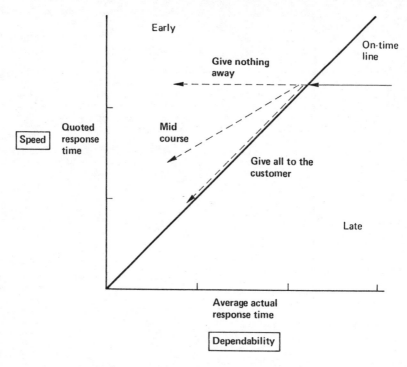

Figure 4.4: Decide on an improvement strategy between 'giving all' and 'giving nothing' away to the customer

- Make delivery dependability a key measure of operations performance, but be sure the way you are measuring dependability really does reflect the type of performance you are after.
- Manage customer expectations by emphasising delivery integrity, and customer perceptions by capitalising on good service.
- Don't underestimate the benefits of stability which internal dependability brings. It is the foundation of most operations improvement.
- Treat the improvement of internal dependability as a means to an end. It shortens throughput time, reduces the need for inventories and makes for punctual deliveries.
- Invest in the care of process technology. The benefits are far greater than traditionally supposed.
- Plan ahead to predict things which might disrupt the operation – at least far enough ahead for you to do something about it.
- Overloading capacity merely causes confusion. Apply strict order input control.
- Consider where flexibility could localise any disruptions.
- Get all departments to take some responsibility for developing their own internal suppliers.

5

CHANGING WHAT YOU DO – THE FLEXIBILITY ADVANTAGE

Flexibility has become one of the most fashionable of manufacturing virtues. Turbulent markets, fast-moving competitors and rapid developments in technology have all forced manufacturing management to reassess its ability to change what it does and how it does it. And that is what flexibility is – the ability to change, to do something different.

Versatility has Virtues

There is no shortage of reasons why operations want to be flexible.

- To cope effectively with a wide range of existing parts, components or products
- To adapt products to the specific requirements of customers
- To adjust output levels to be able cope with demand variations, such as seasonality
- To expedite priority orders through the plant
- To cope with plant breakdowns
- To provide adjustments in capacity when demand is very different from forecast
- To cope with the failure of suppliers (internal and external)
- So that future generations of products can be manufactured on the same plant
- Because there is no clear idea about how much capacity will be needed in the future
- Because there isn't any accepted forecast or plan for the future, so options need to be kept open.

The first four reasons are about flexibility being used to cope with the *variety* of activities that manufacturing has to deal with in its (short-term) day to day operations – variety of products, variety of output levels or variety of delivery promises.

In the next three reasons flexibility is wanted because it maintains performance under *short-term uncertainty* – coping when dependability is poor and things do not go to plan.

The next two reasons relate to the *long-term uncertainty* inherent in all considerations of future strategy. The idea is to put enough flexibility into the total manufacturing operation to cope with whatever conditions emerge – new products, new markets, or new competitors.

And so it might seem for the final reason, but not so. The difference here is that the manufacturing function has no clear idea of what it might be expected to do, even in the relatively near future. It needs to be flexible only because of its *ignorance* of the other functions' plans, or total company strategy, assuming that either exist.

Not all these reasons are equally valid, of course – machines shouldn't break down and deliveries ought to arrive on-time. If flexibility is used to compensate for a lack of dependability it is, to some extent, wasted (see Chapter 4). Better to tackle the lack of dependability directly; but in the meantime flexibility can help. More worrying is the final justification for being flexible – ignorance of what is, or will be, required of the operation. Again, ideally it should never happen. Any functional strategy should follow from overall competitive strategy, and be put together in consultation with the other functions. An operation which is flexible enough to fit in with strategic direction no matter what it is, at best will be using its capabilities in a hopelessly ineffective manner. The whole point of manufacturing strategy is to focus clearly on a specific set of objectives.

Yet it is not only in struggling companies where manufacturing policies are formulated in a partial strategic vacuum. The fact is that most functional strategies need some 'strategic flexibility' built in to their plans to cope with possible changes in direction by the rest of the organisation. It may be regrettable but it can't be ignored.

Flexibility is the operation's shock absorber   Manufacturing needs to be flexible because it has to manage the operation under conditions of variety, long and short term uncertainty and ignorance. The right level of flexibility lets the operation get on with its job in spite of these things. It gives protection against externally and internally generated shocks. And that is how to think of flexibility: as the operation's shock absorbers (see Figure 5.1.) A flexible operation is one which maintains and improves its

78

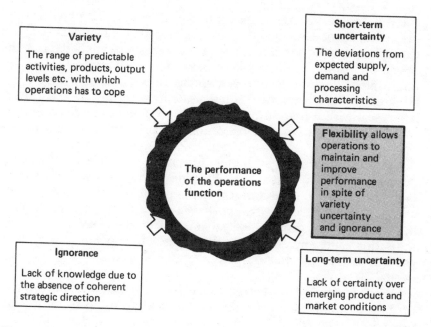

Variety

The range of predictable activities, products, output levels etc. with which operations has to cope

Short-term uncertainty

The deviations from expected supply, demand and processing characteristics

The performance of the operations function

Flexibility allows operations to maintain and improve performance in spite of variety uncertainty and ignorance

Ignorance

Lack of knowledge due to the absence of coherent strategic direction

Long-term uncertainty

Lack of certainty over emerging product and market conditions

Figure 5.1: Factors affecting the performance of the operations function

performance in spite of the impact and jolts from an uncertain environment and the wide variety of conditions under which it has to operate.

Flexibility is only a means to other ends In fact flexibility has a rather special role within the set of five performance objectives. Unlike quality, speed, dependability and cost, it is not always an end in itself – something one competes on directly – but a means to other ends. There is little intrinsic merit in flexibility for its own sake. Rather, operations need to be flexible so they can improve some other aspect of their performance. Companies rarely sell flexibility, they sell what a flexible manufacturing function can give. In other words the justification for enhancing flexibility is usually instrumental, and indeed often expressed in terms of other attributes of manufacturing performance. Look at it as a 'second order' manufacturing objective, but none the less important for that, it has a profound effect on other aspects of performance. The implication being that before deciding how to be flexible it is more than usually important to think through why you want to be flexible.

Flexibility gives better dependability, costs and speed Dependability is enhanced by a flexible operation because it helps to cope with unexpected interruptions in supply. Problems over deliveries, process capability or labour can be localised. As examples: a wide range of in-house processing capability lets the operation produce internally if vendor response is slow or interrupted; labour flexibility which allows transfer of people between departments can compensate for temporary shortages; processes with broad capability can accommodate products re-routed from a part of the operation suffering breakdown, and so on.

Costs are improved by better utilisation of process technology, labour, or material resources. Flexible operations directly contribute to this by overcoming such problems as long process changeover times, excessive work in progress, fluctuating demand between product groups and so on. All of these reduce resource utilisation and therefore increase costs. In addition fast changeovers reduce batch sizes and therefore inventory and working capital.

Speed, meaning fast delivery, fast development of new products, or fast customising of products can be improved by a flexible operation. Flexible changeovers give small batches and fast throughput, and processes with a wide range of capabilities can accommodate new products without costly and time consuming new investment.

Box 5.1

Flexibility might well be the focus of the competitive battles of the future. One survey[1] reports that an emerging trend is the emphasis which advanced manufacturers place on enhancing their flexibility. More especially, they seem to be concerned with attempting to overcome the trade-off between flexibility and cost-efficiency. The period 1975–85 could be called the era where manufacturers discovered that quality and cost-efficiency were not necessarily opposing objectives, the period 1985 through the 1990s could well be the era in which manufacturers draw similar conclusions about the relationship between flexibility and cost-efficiency. In particular, Japanese manufacturers are reported as leading in this change of view. Look at the ten most important action plans in Europe, North America and Japan, as revealed by the survey.

EUROPE	NORTH AMERICA	JAPAN
Direct labour motivation	Statistical process control	Flexible manufacturing systems
Production and inventory control systems	Zero defects	Quality circles
Automating jobs	Vendor quality	Production and inventory control systems
Integrating infomation systems in manufacturing	Improving new product introduction capability	Automating jobs
Supervisor training	Production and inventory control systems	Lead-time reduction
Manufacturing reoganisation	Statistical product control	Introduction of new processes for new products
Integrating information systems across functions	Integration of information systems in manufacturing	Reducing set-up times
Defining a manufacturing strategy	Developing new processes for new products	Direct labour motivation
Lead-time reduction	Direct labour motivation	Worker safety
Vendor quality	Lead-time reduction	Giving workers a broader range of tasks

Note the far greater emphasis placed by Japanese firms on flexibility and related plans.

THERE ARE DIFFERENT TYPES OF FLEXIBILITY

Flexibility is not a simple single aspect of manufacturing performance. It is in fact a shorthand for several quite different attributes. For the sake of simplicity we have not yet differentiated

THE MANUFACTURING ADVANTAGE

one type of flexibility from the another, but it can be a mistake because there are different types and dimensions, and it is important to distinguish between them.

The word 'flexibility' means two different things. One dictionary definition has flexibility meaning 'the ability to be bent'. It is a useful concept which translates into operational terms as the ability to take up different positions or do different things. So one manufacturing system is more flexible than another if it can do more things – exhibit a wide *range* of capabilities. For example, it might be able to produce a greater variety of products, or operate at different output levels.

Yet the range of things an operation can do does not totally describe its flexibility. The same word is also used to mean the ease with which it can move from doing one thing to doing another. An operation which moves quickly, smoothly and cheaply from doing one thing to doing another should be considered more flexible than one which can only achieve the same change at greater cost and/or organisational disruption. Both the cost and time of making a change are the 'friction' elements of flexibility. They define the *response* of the system – the difficulty of making a change. In fact for most types of flexibility time is a good indicator of cost and disruption, so response flexibility can usually be measured in terms of time.

So the first distinction to make is between *Range Flexibility – how far* the operation can be changed; and *Response Flexibility – how fast* the operation can be changed.

Nowhere is this distinction more important than in judging the flexibility of process technology. Flexibility, after all, is supposed to be a major benefit of much Advanced Manufacturing Technologies (AMTs). Robots, Flexible Manufacturing Systems (FMS), Computer Aided Design and Manufacture (CAD/CAM) are all supposed to enhance an operation's flexibility. But do they? And if they do, what kind of flexibility?

Largely it depends with what the various types of AMT are being compared. (Chapter 7 describes the various types of AMT.) For example if an FMS is being used to manufacture what was always a relatively narrow range of parts such as engine blocks (see Box 7.2 in Chapter 7 for an example of this), the most likely alternative technology would be a transfer line. The FMS would certainly be considered more flexible in a number of ways. Its range flexibility would be superior both in the short term (changing from one engine block configuration to another) and the long term (making an entirely different prismatic product).

Likewise its response flexibility would be far superior to the transfer line. Changing from machining one block configuration to another might take no more time than machining two identical blocks.

But what if AMT is being approached from the other direction? Take for example an FMS installed by an aerospace company who had previously relied on a machine shop of conventional and NC machine tools to manufacture a wide range of parts. Here the 'flexible' manufacturing system turned out to be far less *range* flexible than the eminently versatile machine shop which it had partly replaced.

The FMS could cope with only a tiny fraction of the huge variety of parts which the machine shop had processed. But within this limited range it did have more *response* flexibility than the old machine shop – it could change quickly between different parts. This gave it considerable advantages in terms of putting parts through the system quickly.

The old machine shop had so much complexity that its high work in progress meant poor throughput speed. This FMS was in many ways less flexible than the original system – but it was far more productive within its narrower range.

The next distinction to make is between the way we describe the flexibility of the whole operation (its system flexibility) and the flexibility of the individual resources which together, make up the system (its resource flexibility).

System flexibility is best visualised by treating the total operation as a 'black box' and considering the types of flexibility which would contribute to its competitiveness. These are:

New product flexibility: the ability to introduce and produce novel products or to modify existing ones.

Mix flexibility: the ability to change the variety of products being made by the operation within a given time period.

Volume flexibility: the ability to change the level of the operation's aggregated output;

Delivery flexibility: the ability to change planned or assumed delivery dates.

Each of these types of system flexibility has its range and response components, described in Figure 5.2 overleaf.

Whatever system flexibility the manufacturing operation wants, it gets it directly from the flexibility of its individual resources. So

System flexibility type	Range flexibility	Response flexibility
Product flexibility	The range of products which the company has the design, purchasing and manufacturing capability to produce.	The time necessary to develop or modify the product and processes to the point where regular production can start.
Mix flexibility	The range of products which the company can produce within a given time period.	The time necessary to adjust the mix of products being manufactured.
Volume flexibility	The absolute level of aggregated output which the company can achieve for a given product mix.	The time taken to change the aggregated level of output.
Delivery flexibility	The extent to which delivery dates can be brought forward.	The time taken to reorganise the manufacturing system so as to replan for the new delivery date.

Figure 5.2: The range and response dimensions of the four system flexibility types

whatever system flexibilities are needed to compete effectively should dictate the type of resource flexibilities it will need to develop.

Resource flexibility means the ability change inherent in

The process technology of the operation;
The human resources who staff the operations;
The supply networks, the systems which supply and control the operation.

For example, a company which needs a high degree of new product flexibility, will require a process technology with enough range flexibility to cope with whatever products it chooses to develop (see Box 5.2). A company with a wide product range needs the mix flexibility to support it, given in part by fast changeovers (response flexibility) in its process technology. Volume flexibility might rely on the operation's ability to change its staffing levels quickly (one aspect of labour flexibility). A company whose customers habitually change requested delivery times may want high levels of delivery flexibility, which in turn requires a flexible supply network. The details will vary with competitive circumstances, but the principle is the same. Let the flexibilities of company resources be developed to fit in with whatever system level flexibilities are required from the total operation (see Figure 5.3).

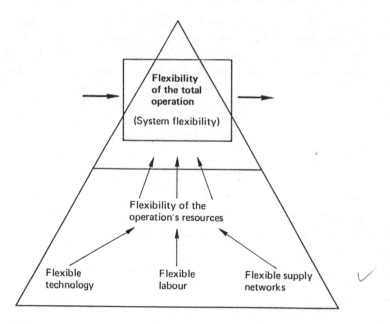

Figure 5.3: An operation's flexibility depends on the flexibility of its resources

Box 5.2

In 1990 Nissan demonstrated how far it had gone towards achieving long-term flexibility in its body assembly.[2] Like most car makers Nissan use robots to weld body panels together. This might sound like a technology flexible enough to cope with changes in product shape and configuration, and so it is – up to a point. Robots often need to be reconfigured and reprogrammed each time a new product is introduced. Also the mechanical wear in the parts makes the robots less accurate as they get older, making reprogramming itself a less straightforward task. New product flexibility is not as easy as it should be, given the accelerating pace of product change.

The company's way of easing product driven changes was its 'Intelligent Body Assembly System' (IBAS), installed initially at its Tochigi and Zama plants near Tokyo. The system consists of some 50 robot jigs and 30 sensors in the 'body in white' assembly part of the production process. It lines up and welds together the car shell's eight basic body panels before sending the body off to the next process. The clever part is that the system's control unit uses the information gleaned from the last attempt at assembling the shell to adjust the jigs for the next shell. The technology learns and adapts – vital in the early days of a new product.

Even more significant perhaps, is that by integrating all the system's robots through a large computer system, they do not need to be physically reconfigured before starting a new model. The machinery can be 'retooled' simply by changing the software. Retooling its body assembly for a new product used to take 11 months and cost 4 billion Yen. Nissan says that IBAS allows them to do it in a quarter of the time and one third of the cost. What is more, the software can be transferred between plants, which allows the company to decide at short notice to transfer the production of a model from one factory to another.

Long-term or short-term flexibility

Without any constraints of time or cost almost any degree of change is theoretically possible. One could merely disband one system and set up another. But this is not flexibility, it is fundamental change.

Where flexibility ends and fundamental change begins is to some extent arbitrary, though 'flexibility' is taken to mean changing an operation's activities without substantial and fundamental changes in its physical facilities. But even within this definition there is still a distinction to be drawn between long- and short-term flexibility. This roughly corresponds to whether range or response flexibility is seen as being the more important.

Companies tend to view response and range flexibility as short-term and long-term problems respectively. That is because most pressing issues of flexibility improvement which can reasonably be resolved in the short-term are usually concerned with improving response flexibility – how fast a change could be made. Immediate flexibility problems include such things as machine change-over times, new product lead-times, purchasing lead-times, time to increase output levels, and so on. Issues of range flexibility on the other hand, are usually regarded as being long-term, involving extra or improved resources. For example, the range of products capable of being produced by the operation is usually determined by the capabilities of its processes and its human resources. Changing the range of products which the operation could make means changing or adding to these resources in a longer term sense than simply speeding up response.

IMPROVING FLEXIBILITY

First, consider alternatives

Notwithstanding the many arguments in favour of at least selective enhancements of operational flexibility, it is not appropriate for all operations under all circumstances. Alternatives to flexibility need to be considered. After all, flexibility in any operation implies some kind of redundancy. If a process has the ability to change what it does, it is, by definition, not using the capabilities which it can change to.

This is not to say that an operation with flexible processes, labour or infrastructure will always be less productive than one whose resources are dedicated to a narrower range of tasks. That will depend on the demands placed on the system. But it does mean that flexibility has costs associated with it. It should not be squandered in areas where it is not needed.

In fact, there are three broad 'flexibility avoidance' strategies.

a) *Compete on a non-flexible basis*
Simply abandon most types of system level flexibility and move the company's market stance towards competing on a non-flexible basis. This could involve, for example, adopting strict limits on product range, attempting to stabilise demand fluctuations and discouraging frequent product modifications. Here the strategy is, in effect, a movement or refocusing of the organisation's competitive stance to avoid operational flexibility – though it may shift emphasis towards marketing's flexibility.

This sounds unfashionable, but in moderation it might be worth considering. For example, a manufacturer of industrial consumables had drifted into a *de facto* policy of supplying 'specials' to some customers; not through any policy decision, but because its sales staff sold to a static, and shrinking, customer base whose requirements were diverging. Eventually the company decided that the strain on its operation was too great. Rather than redesigning its operation to cope with the increased variety, it reconfirmed its original policy of standardised products, and refused business which involved any customising. This initially lost some sales and certainly required a far more energetic and radical marketing strategy. But in the end it paid off. The benefits of simplifying the operation outweighed the costs involved in increasing the flexibility of Marketing.

b) *Reduce the need to be flexible*

Reduce the overall need for flexibility throughout the operation, by reducing complexity; much of which is either generated internally or can be reduced by internal policies. For example, a manufacturer of domestic appliances totally redesigned its product range, so as to increase the amount of parts commonality. This not only reduced variety (hence the need for mix flexibility) of parts production, it also allowed the company to delay committing parts to final assembly until late in the operation, so enabling it to adjust for demand uncertainty.

c) *Containment*

Confine the need for flexibility to selected parts of the operation. This usually means restricting variety or uncertainty to a limited part of the plant if possible. It makes sure that flexible resources are used where they are most needed, while at the same time protecting the parts of the system devoted to standardised operations.

The most common way to do this is to segment or compartmentalise the operation in some way. For example, a manufacturer of garden conservatories made standard products for the domestic market and also offered a customised service for larger commercial customers. Although the technology required to manufacture both types of product was almost identical, their market needs and flexibility requirements were very different. The solution was to separate the plant into two more or less independent manufacturing systems. One developed the new produce and mix flexibilities for its industrial customised products. The other concentrated on the domestic products which need volume flexibility to cope with a 'standard' but seasonal market.

Second, clarify manufacturing objectives

Flexibility is only a means to an end, which is enhanced manufacturing performance. So start by understanding the ends before concentrating on the means. And the 'ends' as far as flexibility is concerned are dependable, low cost and speedy manufacture under conditions of variety, uncertainty and (sometimes) ignorance.

So the second step towards more appropriate flexibility should be a well defined set of manufacturing priorities. First question,

'Which is the most important performance objective: dependability, cost or speed?' Second question, 'What types of flexibility would best contribute to enhancing that performance objective?' It's not that these two questions will exactly define what types of flexibility are required but they do form a sound performance related backdrop to that decision.

Third, clarify why you need flexibility

Go back to the four generic justifications for flexible operations – variety of manufacturing tasks, short-term uncertainty, long-term uncertainty, or plain ignorance of strategic direction. Whichever of these is the prime motivation gives further clues to which types of flexibility to develop.

Variety Manufacturing has to cope with variety in a number of forms – a wide range of components or products, operating at a number of output levels, or catering for many differing customer requirements, for example. The essence of 'variety' as it is being used here is that the range of activities are predictable. There are no major surprises in what the manufacturing function is being asked to do. Nevertheless a manufacturing function which has a wider range of, even predictable, product types or output levels to manage is in greater need of flexibility than one with a narrower range.

If variety is the main stimulus of flexibility then the concerns of manufacturing are likely to be directed towards response rather than range flexibility. Response flexibility is the ability to move quickly and economically between activities, just what is required to cope with variety while keeping costs down and customer lead time short.

Short-term uncertainty Not everything in the manufacturing environment is totally reliable. The unexpected, or at least unplanned, does occur and must be accommodated in some way. Supplier problems, plant breakdowns, demand forecast errors all require some degree of flexibility from the manufacturing function if its performance is not to be driven too far off course. (The previous chapter expanded this idea.)

Since the nature of the unexpected is surprise, the most useful response is a quick one. So again response flexibility – the time taken to change manufacturing arrangements – is the main concern when such short-term uncertainty is the stimulus. Here, though, response flexibility is useful because it maintains dependability performance. A manufacturing operation which uses its flexibility, for example to reroute products quickly following a machine breakdown, or organises an extra shift quickly when demand forecasts prove pessimistic, does so to keep its delivery promises despite the unplanned circumstances. But also range flexibility could, at times, be useful to cope with short-term uncertainty. If, for example, supply difficulties force a company to turn to in-house manufacture, a wide enough range of processing capability would be vital.

Long-term uncertainty Beyond a certain point in the future no firm, however in command of its future, can be totally sure of the demands which will be placed on it. The two basic uncertainties are, first, what type of products will need to be made, and second, in what volumes. This indicates the types of flexibility which are likely to be important when long-term uncertainty is the main stimulus – new product and volume flexibility. The former determines the company's ability to design and make the (possible) new future products, the latter its ability to change capacity and/or output levels.

Ignorance It shouldn't be necessary to squander flexibility on coping with ignorance, and in a company which has well-formulated competitive functional strategies it won't be. Here operational flexibility is needed not to cope with the understandably uncertain movements of the market and competitors, but to compensate for the organisation's lack of strategy direction. And who knows what capabilities will be needed, so who knows what flexibilities to plan for? This is the frustratiion when flexibility is being used to compensate for ignorance – it is downright inefficient. The only way forward is to try and define the gap between what the operation will definitely have to do in the future and what it definitely won't have to do. This gap will then define the extent to which flexibility needs to be built into Manufacturing's plans.

Fourth, draw range-response curves

Range and response dimensions of flexibility are related: the more you change, the longer it takes. But a mechanism is needed which can model both the range and response dimension of flexibility and be straightforward enough to compare the views of more than one manager. One such device is the 'range-response curve', which, as its name implies, shows the range or extent of change possible for varying response times.

Box 5.3[3]

When an operation has to cope with both uncertainty and variety at the same time, and production lead-times are long, flexibility is a doubly useful attribute. The Scotch whisky industry is like this. Demand is geared to the economic cycle and influenced by world-wide fashion trends. Variety in most plants includes bottle size from 75 cl. to several litres, many blends and/or ages, and different labels for each country. Most limiting, the product has to stay in the barrel for at least 3 years before it can be sold as Scotch. After some difficult years and a whole series of mergers the industry is looking towards cutting its costs, reducing work in progress and finished stocks and increasing it responsiveness. The route towards the goals is seen as improving operational flexibility.

Whyte and Mackay is a good example. On the bottle line, it used to take two engineers most of a day to swap the line from (say) a 75 cl to a 1 litre bottle. Now it has reduced the time to 50 minutes, with the line back to peak filling efficiency within one and a half hours. They now change over the line four or five times a week, far more often than before. The result is predictable – a significant reduction in finished stock. Similarly with the 'dry' ingredients (which account for about 25 per cent of the pretax cost of the product). Each of the company's 186 markets used a different label, for example. As well as trying to persuade their suppliers to reduce lead times they rationalised design to allow one basic label to be overprinted locally for, say, twenty different markets. The response time for local overprinting is around three days, so stocks of labels have fallen.

Demand uncertainty is more of a problem. Most whisky sold is four to six years old, and forecasting sales that far ahead is not easy. Demand rarely is exactly as forecast and the industry is obliged to flex its products' specifications to utilise its stocks. The law specifies the minimum proportions and ages of the constituent whiskies in a blend, but it doesn't prevent the company raising the specification by including older whiskies. Alternatively if sales are below its supply capabilities it can keep the whisky until the next 'marketing year'. The price it pays though is not to be ignored – ever year 2 per cent evaporates!

Take for example the volume flexibility of two plants as modelled in Figure 5.4. Plant A finds it relatively easy to increase output by 10 per cent in a few days through overtime, but after this point can only increase it by putting on an extra shift up to a limit of 50 per cent when the physical capacity of the plant is reached. Plant B on the other hand has surplus capacity on its present shift system but needs some time to organise an increase in material inputs and so can increase output only after a delay. At that point output can be increased by almost 20 per cent, after which relatively little more output can be squeezed out of the plant before its capacity limit is reached.

Range response curves can be useful devices during flexibility audits, yet they do need to be adapted to circumstances. For example, range flexibility might not be even indirectly quantifiable; in which case descriptive rankings can be used. Take the case of an electronic systems manufacturer auditing their product flexibility. (Figure 5.5.) When examining the range of new products which might have to be developed, their ranking (in increasing order of difficulty) went:

- customised variants of existing products
- novel arrangements of existing modules

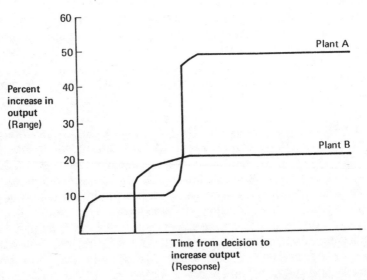

Figure 5.4: Range-response curves for the volume flexibility of two plants

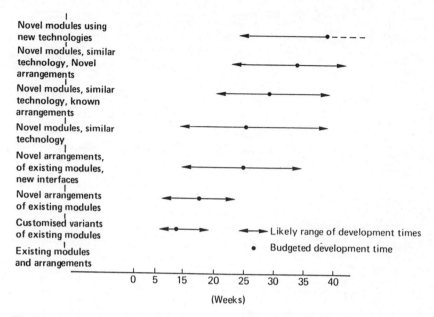

Figure 5.5: Range-response curve for product flexibility in an electronics company

- novel arrangements of existing modules requiring new interface development
- novel modules using similar technology
- novel modules using similar technology in a known arrangement
- novel modules using similar technology in novel arrangement
- novel modules using different technology

and so on.

The major value of range-response curves is that they provide a focus of debate within the organisation as to the likely nature of what is a rather uncertain and fuzzy concept. Total agreement as to the exact shape of the curve for a particular type of system flexibility might never be achieved. Some kind of form, however, must be found which gives the general idea of system capability.

Fifth, develop flexible resources

So far we have dealt with flexibility mainly at the systems level, but different system flexibility requirements imply different operational resources, or at least different ways of organising the

93

Resources	Product flexibility	Mix flexibility	Volume flexibility	Delivery flexibility
Process technology	Range of process capability Capability of design technology	Range of process capability Process change times Scale and integration of processes	Total process capacity Speed with which processes can be focused on required product range	Total process capacity Speed with which processes can be focused on required product range
Human resources	Range of design skills Range of process skills Transferability of labour	Range of process skills Direct/indirect task transferability	Overtime capability Transferability of labour	Overtime capability Transferability of labour
Supply networks	Supply of design and process labour Ability to modify process technology Project management skills	Purchased items lead times Rescheduling capability	Ability to recruit new and/or temporary labour Ability to organise subcontract supply Order processing and forecasting sensitivity	Purchased items lead times Ability to recruit new/temporary labour Ability to reschedule activities

Figure 5.6: *Resource implications of system flexibility types*

resources. Again, it is useful to differentiate between the contribution to flexibility which can be gained from the three activity areas – process technology, the development and organisation of human resources, and supply networks. The table in Figure 5.6 gives a brief guide to the more important resource characteristics that are required for each type of system flexibility.[4] Chapters 7, 8 and 9 go into more details.

Box 5.4
Of all aspects of flexibility, probably the most dramatic improvments have come from changeover time reduction. Machine changeover times which were once seen as a given attribute of the technology itself have been slashed in all types of manufacturing. The following are typical examples.

- GEC Electric Motors (UK), winding machine changeovers down from 70 minutes to 15 minutes.

- Andreas Stihl KS (Germany), heavy stamping presses in power saw manufacture, die changes down from 2 hrs to 3 minutes.
- Alco Lawnmowers (UK), turning and pressed parts, changeovers down to 65 per cent of original time.
- Kodak (Australia), coating technology, line change times down from 16 hours to 2 hours.
- Albion Pressed Metals (UK), again pressing and cropping machine changeovers down from 3 hours to 30 minutes.
- NKK Aluminium Rolling (Japan), hot roll change times down from 2 hours to 10 minutes.
- Cummins Diesel Engines (USA), metal cutting technology of various types, average changeover times down 85 per cent.

There cannot be an industry or a technology where similar success is not possible (although some technologies are clearly easier than others). And why the rush to achieve these levels of response flexibility? Because it is the basic foundation for improvement. All the companies cited above also report equally dramatic improvements in throughput times, stock turns, manufacturing costs and space utilisation.

PRACTICAL PRESCRIPTIONS

- Operational flexibility is an important and pervasive concept, but also an extremely fuzzy and multidimensional one. Differentiate between the different types and dimensions of flexibility.
- Sort out why flexibility is needed. Is it to cope with variety of products or output levels, short-term uncertainty, long-term uncertainty, or because no strategic direction is evident? Flexibility can act as a 'shock absorber' which protects the operation from these disturbances.
- Flexibility affects three other performance objectives. It can help make the operation more dependable, it speeds up throughput, and it reduces costs directly. Decide if one of these stimuli is more important than the others.
- 'Flexibility' has two meanings – how far the operation can change (range flexibility) and how fast it can change (response

flexibility). Try to specify which is more important.

- Think about the flexibility of the whole *system* under the following four headings,

  Introducing product or service changes – New product flexibility;
  Making different mixes of products – Mix flexibility;
  Adjusting the volume of output – Volume flexibility;
  Changing delivery dates – Delivery flexibility.

- Distinguish between the flexibility of the operations *resources* by thinking in terms of

  Technology flexibility;
  Labour flexibility;
  Supply network flexibility.

- Develop a flexibility policy by working from why flexibility is important to the operation and what performance objectives it is required to maintain. Let these factors define the system flexibility which is needed. This in turn should indicate what kind of resource flexibility is required.

- Don't make life unnecessarily difficult for yourself by generating the need for flexibility internally. It's bad enough that the market demands flexibility without bad design, poor communication, lack of focus, excessive routing complexity, and year-end spurts making things even worse.

6

DOING THINGS CHEAP – THE COST ADVANTAGE

It has been fashionable in some circles to play down the importance of cost as a manufacturing objective. As markets and manufacturers gain in sophistication, it is argued, quality, innovation and customer service are at the forefront of competitiveness, not price. The traditional concerns of manufacturing with efficiency, resource utilisation and cost, should therefore be overthrown and in their place a reordered set of goals can be constructed, which better reflect the new competitive order. But this is, at best, a half truth, an overreaction to the narrow and short-sighted obsession with cost which many manufacturers still exhibit. Low cost manufacture is a legitimate and desirable manufacturing aim, even when competitive success is not primarily a matter of undercutting the competition.

There is no lack of importance implied by dealing with cost issues only after examining quality, speed, dependability and flexibility. On the contrary cost is at the centre of manufacturing objectives as the one attribute which impacts directly on the bottom line. Improving outgoing quality, delivery lead-time, on-time delivery and operational flexibility should eventually make their impact on the revenue line. But the influence of reduced manufacturing costs is immediate and direct. More than this, manufacturing is clearly identified in the corporate consciousness as having responsibility for a significant part of operational costs.

The argument here is not that cost is always the most important manufacturing objective – it isn't unless you compete primarily in price. Nor even that the other manufacturing objectives are unimportant if you do compete on price. Rather it is that although manufacturing objectives should be primarily dictated by competitive priorities (which differ with different competitive circumstances and company strategy), cost performance will be important no matter how you compete. Not only because it may well allow lower prices which may significantly enhance competitiveness but because it can directly increase operating margins.

Cost is only part of return

In the longer term, Manufacturing's contribution to the financial return on all the assets which are invested in the operation is the prime measure of performance. Cost is only a part of this, but Manufacturing's influence on return on investment (ROI) is not limited to reducing cost, important though this is. It has a wider contribution to make. Look at the ratio analysis in Figure 6.1; it is crude but it makes the point.[1]

The simple ROI ratio, profit over total investment, is broken down into 'profit/output' and 'output/total investment'. This first ratio, in effect average profit, can be further broken down into average revenue minus average cost. Manufacturing indirectly affects the former through its ability to deliver superior levels of competitive performance. It affects the latter through the more productive use of its resources. And these two measures have been seen as the great manufacturing balancing act, keeping revenue high through standards of service and competitive price whilst keeping costs low. But what about the other part of the decomposed ROI ratio – output/total investment? This ratio represents the output being produced for the investment being put into the operation. It is shown in Figure 6.1 broken down into three

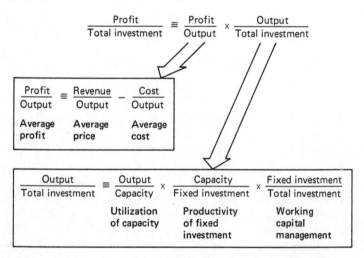

Figure 6.1: Ratio analysis demonstrates that the operations function can influence return on investment in more ways than keeping costs down

ratios, 'output/capacity', 'capacity/fixed investment', and 'fixed investment/total investment'. Take them in turn.

First 'output/capacity', or the utilisation of the operation. This is determined largely by the operation's ability to adjust its capacity to match demand. Operations with a high degree of volume flexibility will be able to keep effective capacity at a level which is close to the demand placed on the operation. Next 'capacity/fixed investment', or what is sometimes called the productivity of fixed investment. This is determined by the skill of the operation's designers and technologists. An operation which achieves the required capacity levels without needing large amounts of capital expenditure will have a better ratio than the operation which has 'thrown money at the problem'. Finally 'fixed investment/total investment', a ratio governed by the working capital requirements of the business. As far as the Manufacturing function is concerned this usually means the level of inventories held.

The point here is that the operations function can influence ROI in more ways than keeping operational costs down – though this is important. It is also important to tackle the other influences, volume flexibility, sharp imaginative and economical design of the process, and tight inventory management.

Attack costs at three levels

Cost cutting exercises have acquired a bad reputation. And it is no wonder. Often they are spasmodic and panic driven reactions to some short-term problem or other which achieve little in real savings, while causing disruption and confusion to the smooth running of the operation. Cost cutting exercises can be ritual activity which gives the appearance of something happening rather than the probably much needed but less comfortable rethink of the underlying causes of cost. Hardly surprising that some say cost should not be tackled head on. It is what they call the 'productivity paradox': the more cost and productivity issues are attacked directly, the less good it seems to do.[2] Rather cost should be tackled indirectly they say. It should be outflanked, improved as a result of improving other things like quality and throughput speed, not by 'cutting costs' as such.

There is considerable sense in this approach but it is only part of the story. Try telling, for example, British Steel that costs will not respond to being cut directly. During the 1980s they transformed themselves from a flabby and loss-making outfit into one of the best steel companies in the world, largely by cutting out surplus resources, and doing so *directly*. Some plants were producing more steel in 1990 than they were in 1980 with less than half as many people. Nothing 'indirect' about British Steel's approach to cost reduction. They demonstrate the first stage in manufacturing cost improvement: get resources balanced and appropriate for the level of output required.

Once resources are at an appropriate level, then is the time for the less direct approaches to reducing cost. The first is to understand the strategic drivers of cost – the decisions which, although taken partly as a response to competitor and market conditions, have a very significant effect on manufacturing cost. The second is to understand the effect the other performance objectives have on cost. Quality, speed, dependability and flexibility can be managed to be the operational drivers of efficiency, not the alternatives to low cost.

Box 6.1

Reducing the number of people in the operation is rarely a painless exercise. It can leave remaining employees dispirited, demotivated, and most important, overloaded. Executed badly, 'downsizing' can disrupt the operation, alienate staff, annoy customers and actually reduce productivity. More than half the companies in one survey reported that productivity either stayed the same or deteriorated after layoffs. Another found that 74 per cent of managers at recently downsized companies said their workers had low morale, feared future cutbacks and distrusted management.[3]

Companies, even those who are aspiring to be 'lean and mean', can be too thin. Cutting head count across the board almost certainly means losing the wrong people. Instead of downsizing, the better approach by some firms is 'rightsizing'. This goes beyond job cutting to a twin attack on methods – cutting the work not the workers; and on unproductive parts of the operation – cut staff only where they are not needed.

Rightsizing means deploying every single person in the operation in the most effective way – being both lean and fit. And it seems to require the following.

- Cut unnecessary work, then staff the work at the appropriate level. Don't just cut jobs out of the operation.
- Keep performance standards paramount. Define exactly what quality, speed, dependability and flexibility is required and do nothing which compromises required performance.

- Change methods – even the most overstaffed operations eventually have to rethink their way of doing things before further cuts can safely be made.
- Communicate what is happening to the operation throughout the operation. When an operation is shrinking its fundamental purpose seems to get overlooked. Tell people what the cuts are for. Tell them how they make for a more competitive operation.
- Take care of the survivors. Fewer people have to be better people. The general practice of good human resource development becomes even more crucial when the operation depends on fewer people.

UNDERSTAND THE STRATEGIC DETERMINANTS OF COST

Manufacturing's ability to achieve low cost production is aided or hampered by a number of decisions which are strategic in the sense that they concern the way in which the company chooses to react to markets and competitors. They can be grouped under three headings:

The *volume* of output for each product group.
The *variety* of products or activities for which the manufacturing function is responsible.
The *variation* in output expected from the operation.

None of these decisions is entirely imposed on manufacturing by the market, or by Marketing, nor can Manufacturing take them alone. Yet they are profoundly important because of the effect they have on manufacturing's cost structure. The very least that is required of the manufacturing function is an understanding of the relationship between these decisions and unit costs.

Costs are Changed by Volume

Volume has always been seen as having an important influence on manufacturing costs. So in some ways it is surprising that many manufacturing managers have little real idea of how costs vary with volume. But they may have a good excuse; the exact relationship between volume and cost is rarely clear.

In the shorter term volume effects are largely a matter of higher throughput spreading the fixed costs of production over a greater number of products produced. In theory the effects of this are straightforward. Figure 6.2 shows how average costs are supposed to reduce as volume increases, according to the formula

$$\text{Average cost} = \frac{\text{total cost}}{\text{output}} = \frac{\text{fixed costs}}{\text{output}} + \frac{\text{variable costs}}{\text{output}}$$

$$= \frac{\text{fixed costs}}{\text{output}} + \text{variable costs/unit}$$

In some types of manufacturing, parts of the process industries for example, cost-volume curves do approach that shown in Figure 6.2. For most manufacturing, though, it is too much of a simplification.

Manufacturing systems accommodate changes in volume in a series of many relatively small discontinuities in the cost curve. Shedding labour from parts of the plant and subcontracting their work when output reduces, for example, or starting up production lines as output increases. Further, nominal capacity is not usually the definite cut-off point implied by Figure 6.2. Capacity is rarely so well balanced between every part of the production process that it all reaches its limit at the same level of output. Bottlenecks

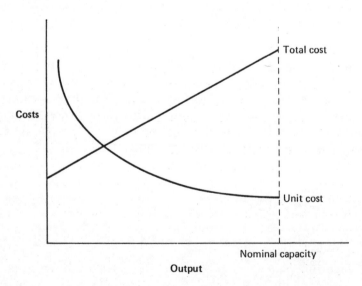

Figure 6.2: The theoretical cost-volume relationship

102

occur as demands are placed on some parts of the plant more heavily than on others. Each part of the plant then has to incur fixed cost steps as it attempts to balance capacity. In this manner output could possibly be raised above the nominal capacity level without large fixed cost hikes. After a certain point, however, the emphasis switches from balancing capacity to the provision of new resources which can mean larger steps in fixed costs – the 'incremental new capacity' zone. Eventually fixed cost steps can become very large, as in the provision of entirely new plants for example. The result is that for most companies the volume-cost curve is neither smooth nor entirely certain, since there is considerable management discretion as to when to commit the plant to fixed cost breaks.

In the longer term the volume of output may allow changes in the way a company uses its technology, or even the acquisition of more economic types of technology. Opportunities in the way technology is used derive from an issue we shall deal with later in this chapter – the effective variety placed on parts of the manufacturing system. It works like this: as volume increases, the tendency is for variety per unit volume to decrease, so each part of the system has fewer different tasks to perform in each time period. At the very least this will reduce the number of changeovers necessary, which will free-up the capacity previously spent changing from one activity to the next, as well as avoiding all the quality problems associated with changeover. More significantly it may allow more dedicated technology to be developed where economies derive from focused specialisation on a narrow set of tasks.

Long-term volume-cost effects also include what economists refer to as 'economies of scale'. Partly these are the economies of bearing the fixed costs of construction only once for a large unit of capacity. Partly, they are the economies gained from the fact that capital costs increase less fast than capacity. In other words a plant's cost will not double when capacity is doubled.[4] Similarly its operating costs are likely to rise slower than capacity. Usually this is because large plants can exploit the benefits of integration more readily.

Big isn't entirely beautiful, though. There are some diseconomies of scale – factors which actually push costs upwards as volume increases. Organisational complexity is one; the larger the organisation the more effort is needed to coordinate its activities. Every marginal addition of a department, unit or section to a plant needs lines of communication with all the existing parts of the organisation. Increasing formality is another factor; organisational structures have a tendency to develop more layers as size increases, which results in less flexibility. The potential for communications being distorted can also increase. And even if effective communication survives, personnel can feel distanced from sources of decision making, the feeling of being 'part of the organisation' fades, and operational improvement become less easy to propagate in a rigid and bureaucratic structure.

All this might imply that there is some optimal balance between the economies and diseconomies of scale. Perhaps there is, but whether it is worth searching for it is less certain. A more productive activity for most manufacturing managers is to understand fully their own economies of scale. Where do fixed cost breaks occur, which parts of the plant are likely to become bottlenecks, at what levels of output do alternative technologies become viable? All are questions which need to be answered before entering into any corporate debate on output or capacity levels.

Costs are Changed by Variety

Variety is the least understood driver of cost. Which is surprising considering that it is an old complaint of manufacturing managers that excessive variety, especially product variety, is a major cause of excessive cost. With high product variety often comes high parts variety, process variety and routing variety. And behind all these types of variety comes the complexity which is the root cause of variety related costs.

High variety first of all requires a more complex technology, or put another way, it makes it more difficult to develop the dedicated technology which keeps costs low. High variety loaded onto plant and equipment usually leads to higher capital and operating costs. Increased complexity of control systems, materials handling and adjustment mechanisms, together with changeover downtimes all contribute to this. But the relationship is not smooth, nor is it static. There are usually 'variety breaks' where an incremental increase in variety cannot be borne by the existing technology, although this is less true for some types of newer process technology, which is breaking down some aspects of the relationship between variety, flexibility and cost. Chapter 7 explores this further.

But it is not technology costs alone which rise with variety. Often more serious and always less quantifiable are the effects of variety on overhead costs. This includes the cost of supporting the increased complexity of the operation. Increased variety of input materials and bought-in parts means more (and a more complex) purchasing effort, records, standards, inventory and coordination. It means more drawings, more product specifications, more process routes and logistics, all of which need organising. It means more finished goods inventory, space, and again, coordination. The overhead consequences of variety are widespread and very significant.

Worse than this, while a good management accounting system may capture some of these variety effects, it can never reflect all the disruptive effects of variety. Every change of the process, the part being manufactured or the route, causes manufacturing management to 'take their eye off the ball', try to cope with the variety, and try to start again. High variety, in effect, makes it harder for the operation to move down the learning curve.

One way of visualising the effects of variety is to start with a Pareto view of the volume-variety relationship.[5] Figure 6.3 shows a typical Pareto curve for an operation. The vertical axis is the cumulative annual sales revenue from all of the products produced by the operation, ranked in descending order of sales revenue. The horizontal axis then represents the cumulative number of different products. Often called the 80:20 curve, the typical finding from plotting such a curve is that 80 per cent of sales revenue is represented by 20 per cent of the product lines.

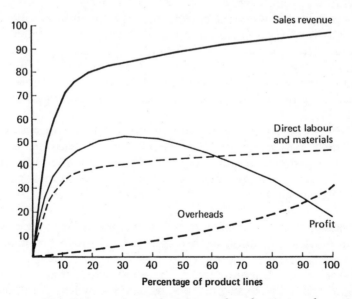

Figure 6.3: The Pareto view of variety related costs and revenue

Plotting operational costs on the same horizontal axis is even more revealing. Direct labour and material costs could well follow a similar pattern to the sales revenue curve. Perhaps in a slightly less pronounced form since scrap rates, yields and labour productivity are all likely to be worse for products which are made infrequently (those represented by the right hand end of the variety axis). Overhead cost however will show quite a different shape. It is at the tail of the variety distribution where the complexity effect of variety makes itself felt. So the smaller volume products represented by the right hand side of the variety axis will often consume a share of overheads far greater than their volume would suggest. The overhead curve in Figure 6.3 shows this effect.

Profit can now be calculated by simply subtracting the direct materials and labour curve together with the overhead curve from sales revenue. It shows cumulative profits peaking at a point typically below 50 per cent of the product range produced. Or put another way, the higher volume lines subsidise the lower volume ones.

Some product cross-subsidy may be inevitable, or even desirable (for example to support new products which, in the long term, will grow to form the high revenue base for future operations). Nor is variety being held up as inevitably a 'bad thing'. It isn't. Many companies compete, largely or partly, on being able to produce a wide range of products. In which case the high volume products (on the left side of the variety axis) are not supporting the products on the right end. Rather it is the low volume lines on the right which enable the high volume lines on the left to be high volume. All products are mutually dependent. Alternatively a company might be able to secure prices for its low volume products which compensate for the increased costs they cause. In which case the tail of the revenue line is raised together with the profit line.

Variety can be a perfectly legitimate part of competitiveness. But the effects of variety *must* be understood. The aim of manufacturing management should not be to reduce product variety necessarily, but to accommodate the maximum required variety in the lowest cost manner. This can be done in a number of ways.

Differentiate between perceived and actual variety Uniform design and standardisation both reduce the effects of product variety. So does a certain amount of redundancy in design. Perhaps the most famous example of this is the Swiss watch manufacturer who produces watches in several designs, each of which is offered either with a date 'window' or without. In fact all watches have a date mechanism. They merely put on a face with a window or without one, depending on the variety being made. It's easier to make just one watch mechanism then add variety later via the face. A wasted date facility perhaps, but it is cheaper than making two different watch mechanisms.

Increase product mix flexibility An operation which can change quickly and cheaply from one product to another is far more resilient against variety than a less flexible operation. This is what was referred to in Chapter 5 as 'mix flexibility'. The dramatic reductions in changeover times described in Chapter 5 are an excellent example of flexibility changing the variety-cost relationship.

Limit variety to part of the plant The resources and infrastructure needed to support high variety manufacture are usually quite different from those needed for high volume, low variety manufacture. So keep the two types of operation apart. This is the idea of manufacturing 'focus' again which was introduced in Chapter 1 and will be treated further in Chapter 7.

Although the effects of variety can be buffered in these ways they can not be eliminated. All operations should at least consider reducing variety in real terms. And while it always seems to be easier to add to a product range than to reduce it, many operations can make some reduction in range without serious loss of business, though if product range is a major element in competitive strategy, it may prove more difficult. One important proviso, however: there is little point in reducing variety if the overhead costs which it spawned are left in place. Reduced variety should mean less support costs. It is the 'appropriate level of resources' argument again.

Costs are Changed by Variation

Variation is the degree to which the demand placed on the whole operation fluctuates over a period of time. Fluctuations in demand come in two types – those you expect and those you don't. Unexpected fluctuations in demand, which take the operation by surprise, have an even worse effect on cost than expected (and therefore hopefully planned for) fluctuations.

The best way to establish the variation-cost relationship is to imagine a perfectly steady demand. All customers are trained to demand exactly the same level and mix of products every week of the year (except for plant shutdown weeks when they don't want anything!). The costs saved under such ideal – and hypothetical – conditions are the costs of variation.

The source of these variation driven costs will depend on how the operation chooses to cope with variation. It might choose to adjust the output level of its operation every time demand fluctuated – the 'chase demand' strategy. Alternatively it might choose to keep output steady and let finished goods inventories absorb the difference between supply from the operation and demand from the market – the 'smoothing with inventory' strategy. The costs of variation will be determined by how these two 'pure' strategies are balanced.

The chase demand strategy Demand chasing means adjusting output to reflect demand fluctuations. It is a strategy which can be achieved in several ways, all of which have cost implications. For example, the quickest and easiest way is to use overtime for increasing output and accept underutilised labour when decreasing output. But overtime means higher total manufacturing costs, although its effect on unit costs depends on the operation's cost structure. And underutilisation, although it means no extra total costs, means higher unit costs because total costs are spread over total output volumes. Often, though, the main argument for using overtime to adjust output on a planned basis is not its cost, but that its speed and ease of implementation usually make it the best 'reserve strategy' for unexpected demand hikes. More permanent changes in labour resources for longer term adjustments in output involve hiring new people or shedding labour. The former requires support overhead costs in advertising, selection, induction, training and so on, as well as the general disruption costs to the operation as new people are assimilated. The latter, as well as the human distress to those directly involved, can mean redundancy costs and the lowering of morale throughout the operation.

An alternative way to adjust ouput might be to subcontract when needed. But again this has its own costs. A policy of purchasing components previously made in-house might bring some benefits (see Chapter 9) but using subcontracting as a buffer to absorb demands on the operation means paying a price in terms of increased overhead support costs and, again, a lack of operational stability. The corollary when downward adjustments in output are needed – taking in third-party work – can be even more costly in the loss of focus it causes in the operation.

The smoothing with inventory strategy The alternative 'pure' strategy to demand chasing is to protect the plant by allowing it to produce at a constant (or convenient) rate and letting finished goods inventory absorb the differences between supply of, and demand for, products. Here the costs involved are outside the plant as such; they derive from the inventory which is acting as the buffer. Working capital costs are the most obvious, but there is the space necessary to house the inventory, and also the risks of holding products before they are demanded – damage, obsolescence and deterioration.

If demand fluctuations are high then the inventory levels needed to absorb the fluctuations will also be high. Under these circumstances this strategy can become excessively costly, especially when demand is uncertain. For example, some cosmetics manufacturers face both very heavy fourth-quarter demand (Christmas) together with uncertain mix and aggregated demand (fashion-dominated market). This makes relying exclusively on the use of inventories too expensive. They rely, as do most companies, on a mixture of the two 'pure' strategies. Although the basic idea holds true no matter which balance of the two strategies is deemed appropriate, coping with variation increases costs.

Market Imperative or Operations Convenience?

High volume then, together with low variety and a steady, predictable demand, keeps manufacturing costs low. Low volume, high variety and fluctuating demand inevitably exact a cost penalty. This does not mean that operations should avoid low volume, high variety, or unpredictable demand. It does not even mean that markets which demand such conditions can only be served at a high operational cost. But it does mean that every company needs to debate how to balance the demands of the market in these respects, against the operational inconvenience and cost they produce.

Manufacturing managers have three contributions to make to this debate. The first is to have a reasonable idea of how costs will respond to changes in volume, variety and variation. Not an exact prediction maybe, but at least an estimate good enough to allow a reasoned judgment to be made. Doing this requires the effects of disruption to the operation to be taken into account as well as the more measurable cost consequences. It also requires identifying the cost breaks – discontinuities in the cost of reducing volume, or increasing variety and variation.

The second contribution of manufacturing is to seek ways of minimising the cost penalties which low volume, high variety and high variation produce. Small scale, low fixed cost technologies – integrated continuous casting mini-mills in the steel industry, for example – can overcome some economy of scale effects.

Similarly modular design principles and flexible operations help to accommodate variety, even if they do not render it entirely painless. The process is a constant search for ways of reducing the cost consequences of volume, variety and variation.

The third responsibility of manufacturing is to make sure that the other functions in the company are playing their part in keeping manufacturing's costs under control. This applies especially to marketing. There are times when manipulating the market to increase volume, or to accept reduced variety, or to flatten demand peaks, is less costly than coping with its effects inside the operation. This is not an incitement to abandon the principle of market-led competitiveness, but it is a plea for exploring the benefits of operations simplicity which a more benignly managed market can bring.

Box 6.2
The breakdown of a factory's full costs varies considerably depending on the type of industry. Some plants bought-in parts and materials accounts for over 80 per cent of their costs, in others the figure is less than 20 per cent. On average though bought-in materials and components represent around 55 per cent of factory costs. Labour also varies from less than 1 per cent to over 40 per cent but an average is around 15 per cent of total costs. Overheads average around 30 per cent.

UNDERSTAND THE OPERATIONAL DETERMINANTS OF COST

Notwithstanding the strategic effect of volume, variety and variation on cost, and the contribution of Marketing in reducing their effect, the prime responsibility for reducing cost still lies within the manufacturing function.

Traditional cost-cutting within the plant tends to concentrate on direct labour, which is puzzling since most operations spend less than 20 per cent of their costs on direct labour (see Box 6.2), and for many types of manufacture it is far less than that. Yet it attracts a totally disproportionate amount of attention. This is what a British Institute of Management survey had to say about it.

It is indeed curious to contemplate the relative managerial effort put into direct labour control and purchasing effectiveness. Whole work study departments are maintained to control the direct labour content of unit cost and, particularly in the engineering industries, vast amounts of management time are put into negotiations on work rating, job allowances and the like. Yet there are many plants that spend twice as much on purchased material as on direct labour that do not even attempt to measure purchasing performance realistically.[6]

A criticism perhaps less true than it used to be, but one which still holds for many operations. Certainly purchasing performance has been neglected. A one per cent improvement in purchasing's bought-in price performance (provided other aspects such as quality and delivery are not compromised) will generally mean two to three times the cost savings of a one per cent improvement in direct labour costs. But what of overheads, that network of cost consuming support which has been called 'the hidden factory'? How can overhead costs be saved?

The answer is that the improvement in this part of cost performance lies in the other performance objectives: quality, speed, dependability and flexibility.

Box 6.3

Why some plants outperform others in their cost and productivity performance is a question with obvious implications. Is it possible to identify the factors which, over a range of different factories, seem to lead to higher productivity not only of labour but of capital and materials? Studies show a reasonable degree of consistency[7] and add support to the idea that cost is best 'outflanked' rather than attacked head on.

Productive plants tend to

1. Have lower work-in-progress
   and the natural consequence of this:
2. Have fast throughput times.
3. Invest in quality programmes
   and a related factor:
4. Report lower waste and scrap figures.
5. Make attempts to balance capacity to avoid bottlenecks in material flow

---

And importantly

6. Try and reduce confusion and establish stability in the operation by

controlling product changes
minimising schedule changes
minimising the expediting of orders
reducing volume fluctuations.

The point here is not that these factors improve productivity. It is that they improve productivity *more than expected*. These factors seem to go beyond their obvious effect. For example, reducing w.i.p. improves productivity more than one would expect if the effect were confined just to the improvement w.i.p. reduction has on working capital.

One issue on which the jury is still out concerns the importance of investment in new technology. There seems to be no automatic connection between new technology and higher productivity of the total operation in the short term. Rather, the effect of investment in technology is likely to affect productivity in the longer term and then only if well managed and integrated with other parts of the plant.

Finally the studies emphasise the importance of overheads in the manufacturing cost structure. Partly because they represent the vast majority of value added in most manufacturing and partly because they provide considerable scope for savings.

Chapter by chapter, as each of the operations performance objectives have been discussed, it has become increasingly clear that error-free quality, speedy throughput, dependable internal delivery, and a flexible operation, are all connected. What is common is that they all contribute to low cost manufacture both directly and indirectly. Summarising from the previous four chapters:

Fast throughput reduces cost Material which moves quickly through an operation spends less time in inventory, attracts fewer overheads and makes forecasting easier; all of which have a positive effect on costs. Fast throughput also encourages dependable delivery since small deviations from schedule can be accommodated faster.

Internal dependability reduces cost Internal dependability reduces the confusion in the operation. If all parts, materials and

112

information are transferred within the operation exactly as planned, the (usually considerable) overhead devoted to chasing up late deliveries is eliminated, as is all the effort of rescheduling to accommodate the late delivery. It also allows throughput to be speeded up. Or, reversing the logic, without internal dependability there is certainly no point in trying to speed up throughput.

Higher quality reduces cost Using 'quality' to mean conformance to specification, the effect on cost becomes clear. Fewer errors within the operation directly reduces rework, scrap and waste. Fewer errors also mean fewer surprises in the operation, more internal dependability and less confusion. Further, error-free operation enhances an operation's ability to reduce throughput time, which in turn reduces costs.

Greater flexibility reduces cost Enhancing some kinds of flexibility can also improve costs both directly and indirectly. Directly by letting the operation change from making one product to another with little loss of output; for example, by increasing changeover flexibility. Indirectly, by reducing throughput time which, in turn, reduces costs. Flexibility can also increase internal dependability, by allowing an alternative process route to bypass a breakdown for example, which in turn reduces cost.

Figure 6.4 shows the general relationship between all the internal performance objectives. The message from it is that internally there should be no trade-off between cost efficiency and the other performance objectives. They all support and reinforce each other. Which, in effect, sets the general question for the operational improvement of cost, 'How can quality, speed,

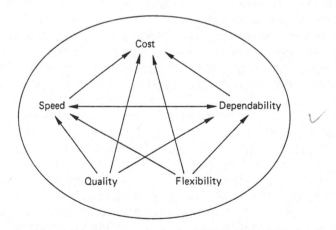

Figure 6.4:   All other performance objectives support cost

dependability and flexibility improvement be made to improve cost performance?'

Box 6.4

Cost is affected by all the other aspects of internal performance. So which do you improve first? Is it better, for example, to start by speeding up throughput, or establishing high quality conformance, or predictable internal delivery, or high flexibility, or what? Alternatively, should initiatives be launched to improve all these things at the same time?

The answer, according to one piece of research which takes its data from a large sample of European manufac- turers, is that there seems to be one particular approach which lays the foundation for sustainable improvement.[8]

First, it says, a precondition to all lasting improvement is an improvement in the *quality* performance of the operation. Only when the operation has reached a minimally accept- able level in quality should it then tackle issues of internal *dependability*. But, and this is vitally important, moving on to include dependability in the improvement process should not stop the operation making further improvements in quality. Indeed improvement in dependability will actually require further improvement in quality.

Once a critical level of dependability is reached, enough to provide some stability in the operation, the next stage is to turn attention to the *speed* at which materials flow through the operation. But again only while continuing to improve quality and dependability further. Soon it will become evident that the most effective way to improve speed is through improvements in response *flexibility*. That is, changing things within the operation faster. For example reacting to new customer requirements quickly, changing production volumes rapidly and introducing new products faster. Again, including flexibility in the improvement process should not divert attention from continuing to work further on quality, dependability and speed. Only now, according to this theory, should *cost* be tackled head on.

The researchers call their theory the 'sandcone model', (see Figure 6.5). The sand is analogous to management effort and resources. To build a stable sandcone a stable foundation of quality improvement must be created. Upon such a foundation one can build layers of dependability,

---

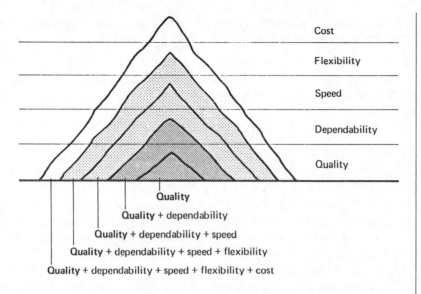

Figure 6.5: The sandcone model of manufacturing improvement; cost reduction relies on a cumulative foundation of improvement in the other performance objectives

speed, flexibility and cost – but only by widening the lower parts of the sandcone as it is built up. Building up improvement is a cumulative process, not a sequential one.

PRACTICAL PRESCRIPTIONS

- Cost is *never* unimportant. Even when competition is based on non-price factors, low manufacturing cost provides the margin to be invested in whatever does count in the market place.
- Remember that cost performance is only part of a more important measure: return on invested assets. Capacity utilisation, economical use of fixed assets and inventory management are also important.
- Approach cost improvement at three levels:

   Getting resources down to an appropriate level to support output.

   Managing the strategic determinants of cost – volume, variety and variation.

   Using the other internal objectives of operational performance – quality, speed, dependability and flexibility – to support cost reductions.

- 'Downsizing' the operation can go too far. Resources should be appropriate for output not inadequate. Think in terms of 'rightsizing' rather than 'downsizing'.
- Rightsizing means

  Cutting out work as well as workers.
  Keeping a sharp eye on operations performance.
  Rethinking methods.
  Keeping everyone informed.
  Taking care of survivors.

- Understand the fixed cost breaks in the volume-cost relationship of the operation. Explore the different ways in which incremental output changes could be accommodated. Define, as far as possible, the affect on costs.
- Understand the importance of product range and variety to competitiveness. Try to compare this to the costs of variety. Especially try to identify the 'variety breaks', the points where marginal changes in variety cause significant changes in costs.
- Reduce the effects of variety through modular design, mix flexibility and limiting variety to part of the plant.
- Balance 'demand change' and 'make-to-stock' strategies to minimise costs.
- Explore ways in which the market can be managed to reduce the need for the operation to cope with volume, variety and variation.
- Look at how costs can be reduced through improving the other performance objectives of the plant – quality, speed, dependability and flexibility.

7

MANAGING PROCESS TECHNOLOGY

More than any other element in manufacturing, process technology defines the nature of an operation. Walk round a garment
manufacturer, a car plant, a frozen food operation, a defence
electronics company, a steel rolling mill, a precision toolmaker,
they all will look, sound, and probably smell, different. It is their
technologies which make them appear as they do. Some technologies are more sophisticated, some linked together, some capable of
doing more things, some just bigger and more intrusive.

SIZE, AUTOMATION AND INTEGRATION – THE THREE DIMENSIONS OF TECHNOLOGY

Even in the most dominant types of technology there is always
some degree of choice; some way in which the technology can be
configured which is different from how it is done now. Exploring
process technology options is a matter of thinking on three
dimensions.

- Its *size* – not its physical size but the scale of its capacity.
- Its *degree of automation* – the balance between capital and labour intensity.
- Its degree of *integration* – how connected it is with other pieces of equipment.

The Scale of Capacity Increment

Some process technologies, such as petrochemical refining, food
processing or steel making, benefit from scale and so tend to come
in large capacity increments. Others, like some metal cutting

technologies, for example spark erosion, are to some extent limited in scale. But no matter what the technology, there is usually some discretion as to how large a piece of plant it would be wise to acquire.

Large machines have several advantages, not the least of which are their economies of scale. Broadly speaking the larger the machine, the less its capital cost per unit of capacity. The cost of the technology itself and the costs of installing and supporting the plant are likely to be lower per unit of output. Similarly operating (as opposed to capital) costs per unit are often lower on larger machines, the fixed costs of operating the plant being spread over a higher volume. But if cost factors are a strength of larger pieces of plant, nimbleness and flexibility can be virtues of smaller scale technology. Mix flexibility especially is enhanced. Four small machines can between them produce four different products simultaneously (albeit slowly) whereas a single large machine with four times the output can produce one product four times faster. Smaller machines can mean lower inventory and more continuous production. They are also more volume flexible. Figure 7.1 shows the forecast volume growth of a product. Large capacity increments will not allow for as smooth a fit between demand and capacity as do smaller capacity increments. This means that with large capacity increments there will be periods of excess capacity, resulting in underutilisation and higher unit costs although, under some circumstances, excess capacity can be useful, especially when it allows the operation to respond to unexpected demand surges.

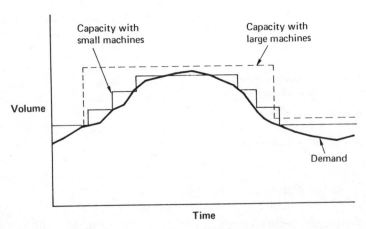

Figure 7.1: Smaller capacity increments allow for a smoother fit between demand and capacity

118

There are two further advantages of smaller machines. First the operation will be more robust; suppose the choice is between three small machines and two larger ones, if one machine breaks down a third of capacity is taken out in the first instance but in the second capacity is halved. Second, it is easier to take advantage of technology improvements. Investing in excess capacity with the idea of anticipating future demands makes the possibility of incorporating future improvements in the technology much more difficult. Alternatively, a smaller machine just equal to current needs can be supplemented by the latest technology when demand warrants it.

Box 7.1

Too many of the operation's eggs in one technological basket can lock companies into a cycle of lost opportunities as well as turning normal strategy-led decision making on its head. Richard Schonberger[1] calls it the 'supermachine cycle'. It goes something like this:

1. Marketing predicts demand growth.
2. The decision is made to increase capacity.
3. A large machine, capable of meeting demand for some time in the future is chosen, on the grounds of its nominally more efficient capital and operating performance.
4. The machine is installed and debugged; a process which is prolonged because of its size and complexity.
5. The machine is underutilised in its early life, resulting in higher than expected costs and longer than expected payback.
6. If demand is lower than expected during this period, pressure grows on marketing to get extra volume any way it can, irrespective of its market development plans. A large costly machine just can't be allowed to stand idle.
7. As demand approaches and passes the machine's capacity, two, then three, shifts a day are worked.
8. Running the machine for such long periods after years of under use, together with insufficient time for maintenance, reduces the effective capacity below the level of demand.

---

9. The case for another 'supermachine' looks good; put
    the capital request in and let the cycle start again.

An example exaggerated to make the point perhaps, but
the point is valid. Technology is dictating policy rather than
being a servant of it. The supermachine is like 'the lion in
the zoo. It demands to be fed on a regular basis, and it eats
a lot'.

The Degree of Automation

No technology is completely stand-alone. All plant and equipment
needs some kind of human intervention some of the time. It may
be minimal, as are the occasional maintenance interventions in a
petrochemical refinery, or the operator may be the entire 'brains'
of the process as in a precision lathe. The degree of automation of
a machine is really the ratio of machine to human effort it
employs, in other words its capital intensity.

Automation gives more than cost benefits

Two benefits of increasing automation are usually cited. It saves
direct labour costs and it reduces variability in the manufacturing
system. Automation is usually justified on the former, but it is
often the latter which is more significant. Nevertheless it is worth
examining what type of labour can be saved through automation
in any particular case. Direct labour can certainly be saved, but
that does not mean that the net effect is an overall saving.

Consider the following points before automating for cost savings
alone.

* Can the technology perform the task better or safer than a
  human? Not just faster (although this can obviously be
  important) but better in a broader sense. Can the technology
  make fewer mistakes, change over from one task to the next
  faster and more reliably, respond to breakdowns effectively?
* What support does the technology need in order to function
  effectively? What will be the effect on indirect costs? Not just
  the extra people and skills which might be necessary but also
  the effect of increased variety.
* Can the technology cope with new product possibilities as
  effectively as less automated alternatives? – what Chapter 5

called long-term uncertainty. A difficult question – who knows exactly what will be produced in the future – but an important one. Automation is a risk as well as an opportunity.

- What is the potential for human creativity and problem solving to improve the machines performance? Is it worth getting rid of human potential along with its cost? Again a difficult but not an impossible question. Some processes have relatively little scope for improvement no matter how creative its operators are; whereas others have considerable potential.

The other advantage of automation is the ability it has to regulate and standardise flow, to make the operation more predictable. To many managers, whose main experience of automated technology is of broken down equipment covered with maintenance engineers, the idea of increased predictability may seem somewhat idealised. But automation at least has the potential to mean more predictable output.

Capital intensive technologies can provide the potential for improved performance. But there is a danger. Technology can be seen as a panacea for all the operation's ills – a 'technology fix' syndrome which avoids the more fundamental problems. Throwing money at the problem is not good management, merely a substitute for it.

Those companies who have attempted the difficult task of separating the benefits which come directly from investment in technology and those which come from methodology and better managerial understanding, have reported some surprising results. Paradoxically, capital investment often makes it necessary to consider the organisation of the operation as a whole, which in turn prompts improvement which is independent of the technology for which it is preparing the way. Look, for example, at these extracts from surveys of FMS installations.

> . . . on average 40 per cent of the benefits predicted for an FMS are in fact achievable, or have been achieved, before the FMS is delivered. This is because the planning process itself has highlighted existing custom and practice which .... can be put right without major investment.[2]

> (We) frequently received estimates that approximately half of the benefits of FMS were derived from managerial and work-based organisational changes.[3]

> . . . companies who operate FMS, and who have presumably undergone some formal evaluation process, tend to concentrate

their efforts more than other manufacturing firms on improving purchasing management and vendor quality, narrowing product lines or standardising products, improving product quality, shortening lead-times and set-up times and integrating information and control systems.[4]

What this means is not that investing in capital intensive, automated equipment is a waste of time. Rather it means that there are considerable benefits to be gained from rethinking the methodology of manufacturing whether or not the investment in technology is subsequently made. But the order should be, first get the methods straight, only then put technology in where it is needed.

Box 7.2

They were seen as the future of manufacturing at one time, used to populate the factory of the future. But robots soon suffered the backlash from the overinflated expectations which were placed on them. Tales of disappointment and money wasted spread and the robotics market stalled. Yet robotics can be made to work and provide the competitive advantage which they always promised. Black and Decker's UK plant at Spennymoor is proof of that.[5]

This plant, one of four European plants in a world-wide network which produces 65 per cent of the world market for power tools, has become a showpiece user of industrial robots. Many are stand-alone installations performing single tasks, but some are built into flexible cells such as the one which assembles drill gearboxes, or the one which puts together the mini-vacuum cleaner. Not every project was without its problems, but all finished up making a contribution to the operation's overall effectiveness. The company's advice to other users and potential users is down to earth.

First, make sure that you have an operations strategy worked out, and only then look for ways of bringing it about, such as robotics. Too many companies see robots as a solution then look for a problem which could be solved by them. Others are only interested in finding a technique to solve a particular problem, without looking at the wider issues.

Second, justify investment in robotics on their contribution to overall effectiveness, not just efficiency. You fail miserably if you look for high utilisation, they say. Although Black

---

and Decker use conventional financial justification based on return on investment and discounted cash flow, it also pays careful attention to other factors. Robots have been justified on the grounds of:

- Volume flexibility – robot cells can work up to 24 hours a day if necessary, virtually unattended
- Quality – robots build quality into the products through their accuracy and repeatability
- Dependability – flexible automation is seen as providing the stability which means that delivery promises are kept.

Finally, they say, look for similar applications elsewhere. Many companies are aware of the applications of robotics in the auto industry – welding, automated screw down and pick and place. But it is far more valuable to seek applications in operations, and market places, similar to your own. Provided, of course, that the possible applications fit the overall competitive strategy.

The Degree of Technology Integration

Along with scale and automation comes the possibility of integration between previously separated functions – meaning the number of distinct functions contained within a linked piece of plant or a system.

It is increasing integration of the various tasks of manufacturing which charts the progress of the so-called advanced manufacturing technologies (AMTs). Cheap micro processing has triggered increasing capital intensity in the basic function of manufacture – designing, controlling, handling and managing. It also allowed the linkage or integration of these functions. Definitions vary (and waste both ink and time) but broadly the terminology at AMT depends on which tasks are being integrated, as in Figure 7.2.

In its physical sense integration is simply the extent to which manufacturing processes are physically linked. This can be brought about either by investment in machines which can perform several tasks, such as a multihead machining centre, or by linking together previously separated machinery by automated materials handling.

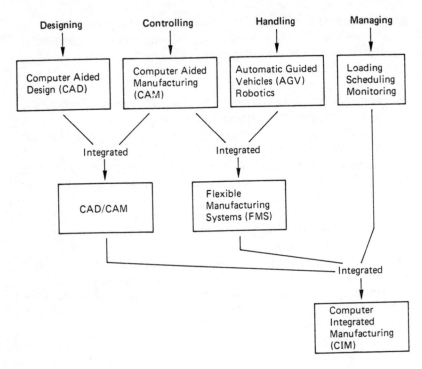

Figure 7.2: Reduced microprocessing costs triggered changes in designing, controlling handling, and managing technologies

The benefits of integration come from the effects of combining several machines into one simple synchronised whole. First there is fast throughput of information and of materials; no inter-machine decision making over which job has priority. Second, and as a consequence of throughput speed, inventory will be lower; it can't accumulate when there are no gaps between operations. Third, flow is simple and predictable; it is easier to keep track of parts when they pass through fewer stages.

But integrated technology tends to be more expensive. Even simple material handling linkages are costly, and if integration involves information systems development (as it does in all the integrated technologies mentioned in Figure 7.2, for example) it is also complex and risky. The more integrated the technology the higher are the skills needed to maintain it. And when it does go down, the whole system is likely to go down. In one sense this makes integrated plant more vulnerable. But if integration involves linking several production stages which would otherwise be loosely connected in series, this 'disadvantage' of integration can be a useful discipline. If one link in a chain stops then the other should do so anyway. To do otherwise would mean producing for work-in-progress, not for the end customer. That is one of the tenets of just-in-time manufacture.

Nevertheless, integration (along with scale and automation) gives proponents of JIT some problems. On one hand integration can link stages together to give synchronisation of flow. On the other it can mean larger, less flexible, more capital intensive equipment. All of which puts an emphasis on technology based progress which fits ill with JIT's preference for simple, continuously improving systems staffed by empowered employees. Not that all JIT disciples are anti-technology as such, but there is a streak of mild Luddism running through some writing on the subject.

Best perhaps to treat integration (and other moves towards 'consolidated' technology) by acknowledging the synchronising effects of integrated technology as the benefit it is. Make sure that wise maintenance practice releases the potential of integrated technology to smooth out variability in flow. But be sure to check out whether the benefits of integration cannot be achieved simply by method, layout and organisational changes. The high-tech solution can seem attractive, especially to the company's process engineers, but why commit to the expense and risks of technology-based solutions if a cheaper and more adaptable alternative is available?[6]

Box 7.3
One of the largest European examples of AMT is BMW's 'Hall 84' installation in Munich. It consists of four large flexible manufacturing cells each of between eight and ten machining centres. Two cells are used to machine the blocks for the V12 engines while the other two machine the cylinder heads for the same engine. The choice to separate the whole installation into smaller cells was taken so that it would give more manageable systems and preserve the flexibility to transfer production of blocks to the cells currently making heads and vice versa.

The total system cost was £40 million ($70 million) which included £3 million ($5.25 million) of software, BMW's largest investment in a single project, although it had previous experience of FMS on its smaller 300 series engine installation. Planning for the project started in April

1984 and was finalised five months later. The machining centres to be used in the system were chosen at the beginning of 1985 together with the overall contractor (Siemens). Installation began in mid-1986 and by January 1987 the first machines were running. Computer controlled production started in September 1987 and the installation was fully operational by the second quarter of 1988. BMW's estimate of payback for the project is around five years of full operation.

BMW say their motives for such an expensive investment were largely influenced by uncertainty over the output rate which it would have to achieve per day. The ability to make other parts on the system gave the company the volume flexibility to increase production volumes for one part without underutilising the plant. In addition the ability to accommodate future design changes and the need for fast throughput to ensure quick response to customer demands was seen as important. The alternative of a series of stand-alone CNC machines was rejected even though it would have given more short-term flexibility. The company felt that multiple handling would have given quality problems, but mainly they wanted to run an unmanned third shift. While a totally unmanned shift was not achieved immediately (they cannot pre-load the machines with enough work for a whole shift) the FMS control can switch production to another machine if necessary to keep production going, which would not be possible with stand-alone machines.

As yet the computer aided design function is not directly linked with the machines without the intervention of production planning engineers, although this was part of the original plan. Communication network difficulties between the functions are blamed for the delay.

Is this experience typical? In a number of important respects yes, it is.

- Advanced technologies are expensive (although this is more expensive than most).
- Software and information technology costs are a significant proportion of the total costs.
- Range flexibilities are not really that great especially when compared with smaller, less integrated, more manual systems.
- Synchronisation, faster throughput, quality and higher utilisation are seen as important advantages.

- So is the long-term flexibility to accommodate design changes.
- Which is probably just as well with payback times of five years (or more in many instances).

Technology influences flexibility

The three dimensions used to define process technology are quite strongly related. The larger the unit of capacity the more likely it is to be capital rather than labour intensive which gives more opportunity for the integration of its various parts.

Flexibility is also related to the three dimensions. Broadly, as scale, capital intensity and integration increase, flexibility decreases. This is because the number of options open to the day-to-day management of the process reduce. There is less complexity, less room for manoeuvre, less ambiguity but also less flexibility. But the relationship of flexibility with the other three dimensions is changing. This is especially true in the technologies 'mid-way' on the other dimensions – those usually associated with engineering batch manufacture. The effect of AMT, and other micro processor based controls, is, at the margin, to give more flexibility for a given position on the three 'dimensions' of technology. So, for example, the FMS described in Box 7.3 is more flexible than the equivalent transfer line but not the alternative machine shop.

Technology influences management's task

Technology choice also has implications for which tasks are key to the effective running of the operation – see Figure 7.3. At the small scale, manual, separated and flexible ends of the technology spectrum, flow of material through the system is intermittent, loading on each part of the (many) parts of the operation changes almost hour by hour. The orders making their way through the complex system could be anywhere so they must be monitored and controlled. At the other end of the dimensions, technology is large, automated, integrated and less flexible. In this type of operation any miscalculations in the choice of the technology will be extremely difficult to overcome. Once the operation is committed to the capacity and nature of its technological structure, the die is cast to a large extent. Moving along the dimensions shifts the whole emphasis of manufacturing management, from controlling on a day-to-day basis towards the long-term design of the plant.

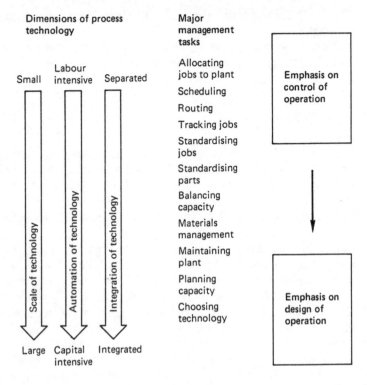

Figure 7.3: Larger, more automated and more integrated techno-logy shifts the emphasis from control to design

Technology must match the product profile

The choice of process technology should be largely determined by the variety and volumes of its products – its product profile. Small scale and separated technologies, because of their flexibility, cope with high product variety more readily than large integrated systems. Conversely the large, capital intensive, integrated techno-logies are ideal to pump out high volumes of standard products where flexibility is of little importance. Figure 7.4 is one version of what is usually called a product-process matrix.[7] The idea is that a manufacturing operation's process technology should be deter-mined by its 'natural' position on the diagonal of the product-process matrix. Thus as a company's product profile changes (as it moves through its product life cycles, for example) so should its technology move through a 'process life cycle'. In other words, it should move along the dimensions of technology in Figure 7.3.

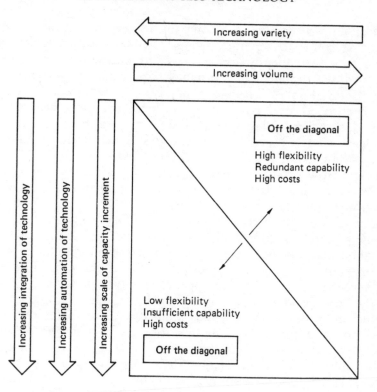

Figure 7.4: The product-process matrix relates the operations product profile to its process technology

Further, there are predictable consequences of deviating from the diagonal on the product-process matrix. Operations with positions to the right of the diagonal have more capability to deal with their product variety than would normally be considered necessary. Such 'surplus' flexibility might be costing the operation by providing redundant capabilities. Operations with positions to the left of the diagonal have less flexibility than one would expect them to have, if they are to cope with their product variety. The consequence of this is higher than expected opportunity costs – the cost of having to shift an inflexible operation with its long changeovers, high inventory or both.

Moving off the diagonal – an example

Yet operations do find themselves off their 'natural' diagonal position. Take for example the case of a manufacturer of hand tools for the consumer market. A trend towards range rationalisa-

THE MANUFACTURING ADVANTAGE

tion of screwdrivers led the company to revise its product ranges replacing some 50 separate product types (many of which were made only infrequently for specialist applications) with a newly designed set of nine. The original large number of types were manufactured in batches on a number of stand-alone processes which cropped steel bar, forged the end, trimmed, heat treated, ground, sometimes plated, marked and inserted the blades into the handles. A set of process technologies which were small scale, fairly manual, not integrated and very flexible.

The narrow product range prompted the company's production engineers to draw up plans for a new production system which integrated several of the operations in the original process sequence. This involved investing in rotary forging, large bed grinders, and induction coil heat treatment, as well as materials handling technology. There would be more capital equipment, fewer people, larger machines and less changeover flexibility. However, a delay in gaining approval for the capital cost resulted in the new product range being made on the old, flexible but inefficient, process. Initially this seemed as if it meant redundant flexibility and high costs. The operation had not moved down the technology dimensions yet the product profile had moved to the right. But rather than accept their cost disadvantages as an inevitable consequence of having inappropriate technology, the company actively tried to exploit the advantages their position gave them. For example the old flexible technology could manufacture smaller batches than seemed to be warranted. But by issuing forecasts weekly, rather than monthly as before, production schedules could be accommodated which matched demand far more closely and gave lower finished goods inventories. The old system was being exploited to improve responsiveness and thus reduce working capital requirements. Admittedly total manufacturing costs were higher than would have been the case with the new technology, but not as high as they would have been without changes to the forecasting and scheduling procedures.

Eventually the new technology superseded the old. But not quite as originally conceived. The company had become impressed with the benefits of a flexible, responsive operation. They were unwilling to sacrifice their changeover flexibility for a more automated, higher production rate, process. Consequently several changes were built in to the new integrated system to allow the same responsiveness as before. In some ways the new system was still less flexible than the old (9 types instead of 50), but not where it mattered.

The lessons here are that, although the product-process matrix gives a general indication of how process technology will differ from differing product profiles, it does not prescribe a 'correct' technology. It gives a general idea of how technology will need to be adapted as product profiles change. But this does not preclude breaking the connection between the dimensions of technology so as to get some of the best of all worlds. Any company finding itself off the diagonal could usefully ask how it can exploit the benefits which its position gives it, and what it must do to overcome the negative effects of its position.

Focus and Segmentation

What then, when the operation's product-process position can not be conveniently located at a single point? For example, suppose a television plant makes three ranges of television. A small screen portable range consisting of only four models – high volume and low margin. A mid range of larger screen models sold both under their own brand name and to the rental companies, where variety, screen size, specification, and brand name were relatively impor-tant and margins varied depending on market conditions. Finally, what was known as 'the rest' within the company – an ever changing mixture of low volume 'end of life' models, experimental batches and very high specification models. Production was characterised by high variety, small volumes and varying, but mainly high, margins.

As far as its product portfolio is concerned, the plant has three quite separate positions, one for each range. But what of its process technology? Should it take some kind of compromise position which could accommodate all three ranges? Or alternat-ively should it segment the plant into separate(ish) parts, each of which could concentrate on its own range and develop its process technology accordingly? This latter policy would allow each part of the plant to concentrate on its own specific task, a concept known as 'focused manufacturing'.

The gist of this idea is that for effective support of competitive strategy the manufacturing function should focus each part of its manufacturing system on a restricted and manageable, set of products, technologies, volumes and markets so as to limit the manufacturing objectives in which it is trying to excel. It simplifies the whole manufacturing task. 'A factory which focuses on a narrow product mix for a particular market niche will out perform the conventional plants which attempt a broader mission . . . focused manufacturing is based on the idea that simplicity, repetition, experience and homogeneity in tasks breed competence'.[8]

The idea of focus has a powerful intuitive attraction. The television manufacturer, like most companies, manufactures a range of products which compete in different ways, possibly in different markets. The whole idea of market segmentation for example encourages different competitive stances for different markets and therefore a corresponding range of manufacturing objectives. A fine idea, but one which can cause considerable disruption to manufacturing and extra costs of machinery and tooling unless it is recognised that the operation itself ideally needs to be segmented to match the parts of the market served.

The clear attraction of focused manufacturing as far as its technology is concerned is that it minimises the number of compromises needed to develop process technologies for separate sets of product profiles. So, for the television plant, the high volume low margin small screen sets need an integrated, capital intensive technology with only enough mix flexibility to cope with its four models. The mid range's process technology needs the flexibility to cope with high variety. Any moves towards integration of the processes or automation to reduce costs can only be pursued if mix flexibility is preserved. The low volume products ('the rest') need relatively general purpose technology to cope with the widely differing specifications within the product group. In effect segmenting the technology builds a separate manufacturing operation for each group of products which needs a separate set of performance objectives – the so called 'plant within a plant' approach.

EVALUATING PROCESS TECHNOLOGY

Whatever disquiet there might have been over financial appraisal methods prior to the age of AMT it was either muted or confined to the furthest reaches of specialist journals. But no longer; the nature of the advanced technologies has exposed the inadequacies of conventional methods of financial evaluation. This is because many of the newer technologies could not be more difficult to handle financially.

They tend to be expensive, so technology choice decisions are not trivial. A large part of a firm's capital budget can be taken up by one or two major projects. The consequences of one piece of equipment failing are correspondingly greater. The technology is intrinsically more risky, more complex, with more to go wrong and closer to the edge of development knowledge. The investment is likely to be 'strategic' in the sense that it directly affects the company's competitive position, and its influence is not confined only to one part of the operation. The benefits of the technology encompass the whole range of competitive objectives, including quality, speed, dependability and flexibility, none of which are easy to translate into financial terms. Yet without incorporating these benefits paybacks become prohibitive.

Such evaluation is both important and difficult. Best to take an approach which distinguishes between,

- the *feasibility* of the investment – how difficult it is to install the technology;
- the *acceptability* of the investment – how much it improves competitiveness and gives a return on the investment;
- the *vulnerability* of the investment – how much risk is involved in terms of what could go wrong.[9]

Figure 7.5 illustrates.

Technology investment must be feasible   If the resources required to install a piece of technology are greater than those which are either available or can be obtained, it is infeasible. Three broad questions are worth asking.

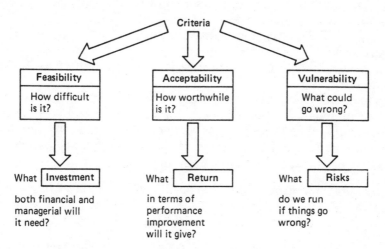

Figure 7.5:   Assess technology by considering the feasibility, the acceptability and the vulnerability of the investment

a) What kinds of skills, technical or human, are required? Every investment in process technology needs a set of specific skills to cope with the implementation. If an investment is similar to the usual activities of the organisation these skills will probably be present. But with a completely novel process, novel implementation skills might be needed.

b) What quantity of operational resources are necessary? This involves determining the number of resources – people, facilities, space, materials, etc. – which would be required to implement the process.

c) What are the funding or cash requirements? For many decisions the major feasibility issue concerns the cash which would be required. For some decisions this could mean simply examining a one-off cost, such as the purchase price. Other, more strategic process investments, may need an examination of its effects on the cash requirements of the whole organisation.

Technology investment must give acceptable benefits A process technology's acceptability is how far it fulfils the company's objectives, in terms of,

> its operational impact, and
> its financial impact.

All process technology should contribute to the business in an operational context. So use operational performance objectives to assess acceptability – giving more weight to those which contribute directly to competitiveness.

Quality: Does the technology reduce the chance of errors occurring in the operation?

Responsiveness: Does the process technology shorten customer's lead-time?

Dependability: Does the process technology give an increased chance of things happening when they are supposed to happen?

Flexibility: Does the process technology increase the flexibility of the operation, either in terms of the range of things which can be done or the speed of changing what can be done?

Assess the financial impact The financial impact of process technology is the comparison of the costs to which the investment commits the operation, and the financial benefit which might accrue.

Ideally both the 'costs' of the investment and the resulting benefits ought to include everything which is influenced by the investment over its life. In fact, this is impossible in any absolute sense. The effects of any large process decision ripple out like waves in a pond, impinging on and influencing many other decisions. Yet it is sensible to include more than those which are immediate and obvious, and a life-cycle approach provides a useful reminder of this.

Conventional financial evaluation has come under criticism in recent years for its inability to include enough relevant factors to give a true picture of complex process investments. The more 'advanced' the technology the more problematic evaluation becomes. With many of its costs and benefits both uncertain and intangible, many complex process investments are justified as an act of 'strategic faith'.

But adapting conventional investment appraisal techniques is more sensible than abandoning them altogether, provided they are used in a manner sensitive to the attributes of advanced process technologies. For example,[10]

- Don't set discount rates too high – it doesn't make for high return projects, it just discourages innovation and competitiveness.
- Evaluate technologies not against current conditions but against the assumption that competitors may invest in similar technologies.
- Don't underestimate the total costs of technologies – software development for example.
- Include all benefits deriving from the technology which can be measured in some way. For example include inventory reductions, reduced floor space and increased quality.
- Take account of the intangible benefits such as increased flexibility, shorter manufacturing times and increased learning. Not necessarily by estimating the financial benefits directly, but by asking what the cash flow from these intangibles would have to be in order to make the investment attractive, then judging whether such cash flows could reasonably occur. For example, suppose the amount of cash flow to bring the return from an investment up to an acceptable level is £300,000 per year for the next 5 years. The question then becomes, 'Do we believe that increased flexibility and other difficult-to-measure benefits will give us an extra cash flow of £300,000 per year for the next five years?' If yes, then the investment is worthwhile.

Assess the Vulnerability of Technology Investment The risk inherent in any process investment is there because one cannot totally predict

a) how it will affect the performance of the whole operation, or
b) the external conditions, prevailing after the investment is made – for example the volume of demand or the interest rate, or
c) the reaction of outside companies to the investment – for example whether competitors are likely to make similar investments.

All need assessing and putting in terms of the downside risk for the operation – the most pessimistic outcome possible. The key question then becomes 'Is the downside risk worth taking?'

A word of warning about assessing the vulnerability of investments in process technology: although it is vital to assess and preferably quantify risks, managers need not take a passive view of uncertainty. Suppose the worst does indeed happen and the new technology runs into problems, the manager is not expected to 'shrug his shoulders and console himself with the thought that even though a bad outcome occurred, he made the correct decision given the information he had at the time.'[11] Resourceful technology management means taking whatever actions are needed to counter the risk or its effects. This will not eliminate the risk or the necessity of including it in the evaluation, but it does change the question from, 'If we choose this technology what would the consequences be if the worst happened?' to 'If we choose this technology, what can we do to reduce the effects of the worst happening?'

LINK PROCESS TECHNOLOGY TO PRODUCT DEVELOPMENT

Process technology shapes the operation. But the product shapes the process technology. Product and process cannot, or should not, be developed apart from each other.

Chapter 3 highlighted the increasingly hectic pace of product technology based innovation. It is the sheer pace of new product introductions as much as the nature of the product innovations themselves which has impacted manufacturing. The way the interface between product and process development is managed is one of the key factors in the successful development of both.

Manufacturing and Product Development are different functions with different roles that employ different types of people. Yet facing up to the barriers which can divide the two functions is a useful starting point for breaking them down.

Time horizons for the two functions, for example, are often very different. Manufacturing's preoccupation is inclined largely towards the short-term. There must be an immediacy about their thinking – it requires a conscious effort to wrench manufacturing's focus up to the longer term. Whereas product development is different. By definition of its function it is looking forward to the next step, or even the one after that. If product development personnel were thinking only about today, they would not be doing their job properly.

The two functions also value different qualities in the individuals they employ. Manufacturing's personnel value a calm mastery of the factory's dynamic complexity, together with fast reactions and problem solving. Imagination and originality are important of course, but not to the detriment of the other qualities. The very opposite applies to product development personnel. Their *raison d'être* is novelty and originality of thought. Without it the development process becomes moribund. Finally there is each function's approach to change. Product development is required to break old thought patterns and challenge assumptions; to embrace change with the fervour of a true radical. Manufacturing on the other had is burdened with the weight of the company's fixed assets. Change is not adopted lightly. The conservative position comes partly as a consequence of its responsibilities and can go some way to explaining manufacturing's resistance to R&D innovations.

The process of taking a product development project from concept through to full production involves several distinct(ish) stages. Terms vary depending on the industry and a particular product development might not include all stages, but typically the stages of product development could be,

1. Concept investigation
2. Basic technology/engineering/formulation investigation
3. Detailed product development
4. Process engineering development
5. Pilot production run
6. Manufacturing ramp-up.

In this example there are six distinct stages of the development, possibly involving different groups of people. There are also five 'mini interfaces' to be managed between the six stages. At each interface there is an *upstream* or previous stage and a *downstream* or subsequent stage.

One of the important decisions in managing each interface is when to commit downstream resources. Should the downstream stage (and the resources needed to execute it) be started only after the upstream task is completed, or when the upstream task is half finished, or even at the start of the upstream task? Figure 7.6 shows four types of relationship: in this example, for the relationships between product and process development.

Sequential where resources for the downstream task are scheduled only after the completion of the upstream task.

Overlap where downstream resources are scheduled to start their task sometime before the completion, but after the start, of the upstream task.

Parallel where upstream and downstream tasks start and finish together. The emphasis is likely to be towards the upstream task at the beginning of the period and towards the downstream task at the end of the period, but the important point is that personnel

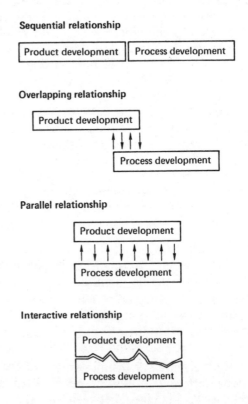

Figure 7.6: The four types of relationship between upstream (product) and downstream (process) development stages

from both up and downstream task teams are assigned throughout the period.

Interactive where not only are personnel from up and downstream teams involved throughout the period, the distinction between the two teams starts to disappear. So, for example, a single team of engineers will work on both the product and process engineering for a new product. Individual engineers might concentrate more on product or process aspects of the task, but their attachment is to the development project itself rather than to a particular stage of its development.

There is a general consensus that the early commitment of resources can be a major factor in increasing the effectiveness of the interface between development stages.[12] Sometimes called 'Simultaneous Engineering', early commitment of resources allows the expertise of the various task teams to be merged at relatively early stages in the design of products. It encourages, for example, manufacturing to make its input into the design process early enough to include manufacturing constraints into the details of product design. This avoids costly changes later in the process, as well as shortening the elapsed time for the two tasks. One of the principal proponents of simultaneous engineering has been the Ford Motor Company, although many other car companies use the concept. Rover, for example, has set up teams which mastermind the whole product development process from concept to production. Its K Series engine was brought to market in record time by using this approach. And also from Rover comes a useful sporting analogy to describe the simultaneous engineering approach. Instead of acting like a team in a relay race, they say, the different groups in the development process should act like a rugby team. Each player should be moving forward in a coordinated manner with all the others.

Box 7.4

Close communication between the various teams involved in the simultaneous engineering of a product is helped by close physical proximity. But it can be made to work even when the teams are thousands of miles apart. Take the development of the LT–5 engine for Chevrolet's ZR–1 Corvette sports car.[13] The engine was developed in four years instead of the previously assumed seven, by using the principles of simultaneous engineering. Yet the three teams who developed the product were General Motors CPC group in Detroit who developed the marketing and product specifications, the Mercury Marine Division in Stillwater, Oklahoma, who manufactured the engine, and Lotus engineering in the UK who engineered it.

One result of using simultaneous engineering was cost savings. For example, it helped to identify problems relatively early in the project rather than after time and money had been spent needlessly. Working as a team also reduced the duplication of the various parts of the total design effort testing for the same problem. Capital expenditure was also handled more efficiently. Mercury could influence the Lotus engineering team to design for their existing equipment in the plant, and where new production machinery had to be bought (well in advance because of long purchase lead times) it chose flexible equipment which could cope with any subsequent design modifications.

Those involved in the project developed what they called the principle of 'flexible teamwork' where, if necessary to save time, team members were expected to go outside their normal task areas. For example in some instances production engineers at Mercury were given product design authority by Lotus. The Mercury engineer could make changes on-site and fax the changes to Lotus for approval later. None of these changes were subsequently reversed.

Where formal approval was required it was kept as simple as possible. Far fewer committees than usual were set up and there were fewer steps in approval processes. Furthermore the approval process changed depending on the stage of development. At the start of the prototype stage the minimum number of signatures were needed for approval. As the design progressed the number of approvals were added and subtracted depending on what was appropriate for the stage of design. During the development of the manufacturing process changes were made quickly by using simple 'Specification Update' forms which could be faxed easily between sites and required only two signatures, one each from design and manufacturing.

According to GM the most important factor which facilitated speedy development was trust. As the various parts of the project gained confidence in each other and as top management's anxieties subsided, the teams were left to do their job without interference, but within time deadline.

PRACTICAL PRESCRIPTIONS

- No matter how obvious process technology decisions seem to be there is always some degree of discretion. And technology choice almost always has significant effect on operations performance. So don't leave the choice to the engineers alone, it's too important.
- Think about alternative process technologies in terms of

  The scale of the technology – how large a machine to buy;
  The capital intensity of the technology – how automated a machine to buy;
  The degree of integration of the technology – how many different processes to have in one machine.

- Don't rush into expensive 'technology fixes'. Are there changes which could be made in the way manufacturing is organised – the 'methodology of manufacturing' – which could give most of the benefits provided by a technological solution?
- The main benefit of many advanced manufacturing technologies is the way flexibility can be retained even with increased scale, automation and integration. Investment in AMT should give either more flexibility for a given scale, automation and integration, or vice versa.
- If there are changes in the volume and variety profile of an operation's product portfolio there should also be changes in the process technology.
- Check where product groups are on the product process matrix. Does it give an indication of whether the operation should be segmented to achieve better focus?
- Evaluate the purchase of process technology by checking its feasibility, acceptability, and vulnerability.
- Adapt conventional capital justification techniques to take account of the uncertainties and intangible benefits of new process technologies.
- The development of process technology should be closely linked with the development of product technology. Use 'simultaneous engineering' principles to manage the interface between the two.

8

DEVELOPMENT AND ORGANISATION

It has always been true that a company's most valuable assets are its people. The irony is that new manufacturing technologies, once seen as displacers of labour, have in fact made this truth more self-evident.

Most technology changes have increased the need for at least part of an operation's human resources to be more skilled in its tasks, more responsible for its performance, and more involved in its improvement.

But it is not only technology changes which have affected the organisation of Manufacturing. The blizzard of new ideas concerning how to control the operation, develop support systems, relate to suppliers, measure performance and contribute to strategic direction generally, have left Manufacturing reassessing its very role in the business.

Not surprising then that amongst these new ideas some should emerge of how the function can develop and organise itself, the better to support the changes.

THE JOB OF MANUFACTURING IS CHANGING

To understand the organisational needs of manufacturing means first understanding the way in which manufacturing activities – the jobs which manufacturing personnel do – are changing. This means looking at the implications of more than technological innovation; it also means examining shifts in the methodology of manufacturing and the way it formulates its strategies. All three are changing, so it is worthwhile to examine the influence each has in shaping the new manufacturing task.

Strategy-making needs new skills

The obvious first point to make regarding manufacturing strategy, is that there is more of it about. Most larger companies now have some kind of formal strategy worked out for their manufacturing function; though some of them arguably are not particularly helpful. Nevertheless manufacturing personnel are increasingly expected to contribute to strategy formulation. Directly by representing the manufacturing function in the debate which shapes the whole organisation's competitive direction, and indirectly by interpreting strategy into its operational implications. This means that as well as having a far broader understanding of strategic business objectives, they need to develop the ability to translate competitive objectives into well-defined action steps. They must be able to cope with the intrinsic uncertainty of market forecasts, and handle the flexibilities which this uncertainty will require. They also must be able to define the objectives of individual parts of the manufacturing system so that they all contribute appropriately to system objectives. Theirs is the responsibility of making the strategic-operational axis of the company work.

Not only do the new manufacturing managers need the internal skills of interpreting and influencing strategy, they also need to look outwards. They need to be, to some extent, the external eyes and ears of the operation. They should know the customers – directly, not by hearsay. They must know the suppliers, again directly. And they must know about competitors. And why should manufacturing managers, at the very centre of the firm, become close to these external influences? Because, to re-emphasise a point made in Chapter 1, they have to bring them into the operation. Not literally maybe (although in the case of customers and suppliers it can be a good idea), but a clear image of customers, suppliers and competitors needs to be well established right on the shop floor.

This is a radical departure from the conventional position of many manufacturing managers, who see the role of the other functions in the business as insulating and protecting them from the outside world. Materials, they believe, come from Purchasing not suppliers. People come from Personnel not the local community. Technology comes from Engineering not vendors. And most dangerous of all, products are manufactured at the behest of Marketing not customers. Surrounded and isolated by the rest of the business, manufacturing operations can not hope to play its central competitive role. The other functions in the organisation

THE MANUFACTURING ADVANTAGE

should be there, partly at least, to facilitate contact between the operation and its environment, not to insulate it.

The factory is also a showcase

Take this point about manufacturing managers becoming more externally oriented to its logical conclusion. Manufacturing managers produce what the customer buys. They serve the customer. They are the direct link between the value adding activities of the firm and its source of revenue. Manufacturing serving customers – it sounds like a supermarket, or a hotel, or a restaurant. And perhaps it should do. Perhaps manufacturing managers ought to think of themselves as service providers. Perhaps customers ought to be encouraged to see the factory as their 'source of service', their supermarket. Perhaps they should be encouraged to come in to their 'source of service' and see their order being processed. Discuss the finer points of its manufacture with the managers who are actually making it. Perhaps the factory ought to be seen as a 'display case', an open, accessible, high-contact operation where the customer should be impressed as much by the way the products are made as by the products themselves.

Now this really is a change. Manufacturing managers showing customers round the plant. Exposing their manufacturing systems to the critical gaze of the customer. No secrets? Nothing which can be hidden? Nowhere to hide?

Or is it such a change? How do operations managers cope with customers? Quite well, actually. And so they should. They talk the same language because they have a lot in common – the effective manufacture and effective delivery of the products. Operations people are talking to operations people. No 'marketing concepts' are getting in the way. No 'segmentation', 'unique selling points', or any other construct is obscuring the fundamentals. Direct contact between producer and customer. There should be a most promising basis within any company's manufacturing management for developing direct links with external 'players'.

And to some extent it does happen naturally. Manufacturing managers do talk to customers and suppliers – but usually to sort out a problem! Complaints, errors, missed deliveries – these are the common currency of most conversations with customers and suppliers. Yet there is scope for a far richer relationship. The factory itself can become its own best advertisement. It can act as a showcase for the company, an opportunity for customers to be impressed by the excellence of their supplier's manufacturing.

144

Manufacturing must not only make things well it must be seen to make things well.

Develop 'high-contact' skills

All this widens the expectations we should have of manufacturing managers.[1] As the walls of the plant become increasingly transparent, they need the skills appropriate to managing a high contact operation – one where customers see at least part of the value being added to their products. Ask the operations managers in the supermarkets or hotels mentioned earlier and they will tell you that customers judge the operation both by its 'products' *and* its processes – by what they get and how it is given. So it follows that effort should be put into the 'image' the operation presents as well as into the product itself. Managing the 'ambience' of a factory may seem far fetched in some industries, but it could start to mean real business opportunities for the company.

Box 8.1

Look at the UK plant of Yamazaki Mazak for an ideal example of how to use a factory as a showroom (and as a schoolroom). The first thing which strikes the visitor (and there have been thousands; the company have never been slow to see the benefits of 'high contact' manufacturing) is the sheer style of the place. 'Bonsai trees and Japanese ornaments adorn airy conference rooms, and a formal Japanese garden overlooks the foyer in this sylvan factory,' as one enthusiastic visitor wrote.[2]

The ostensible motive for Yamazaki's open door policy is to show off their own products. What better way to convince customers to buy machine tools than to show them making other machine tools? But it goes deeper than that. The plant is clean and patently well ordered. The products are displayed in a state-of-the-art configuration. The company's managers will not hesitate to explain how quality and now productivity matches their Japanese plants. The underlying message is 'The products are good because we are good at making them'.

Technology changes need skill changes

It is the sheer speed of change in many manufacturing technologies which is so startling. Some say that the potential of technology in manufacturing is now further ahead of our understanding of how best to exploit it, than at any time in the past.[3] Few aspects of manufacturing management remain unaffected, and the dominant cause, as was discussed in the previous chapter, is the introduction of computers to the manufacturing environment.

The obvious point to make is nevertheless important: far more dependence is going to be placed on an organisation's technical skills and judgments. As companies look towards technology to enhance competitiveness, then the potential contribution of manufacturing engineers becomes increasingly obvious. The potential capability of most computer based, advanced manufacturing systems is considerably greater than the capability which is likely to be needed in the short term. Selecting or designing manufacturing systems is no longer simply a matter of asking whether the system is capable of making what we want, how we want it. The question now should be whether the manufacturing system provides sufficient capability to produce not only what is wanted in the short term but what is likely to be made in the long term. The expanded capability of much manufacturing technology will inevitably increase the time horizon of technological decision making.

The rise of the Manufacturing Systems Engineer

The responsibilities on the engineers who make these decisions are heavy. To shoulder them, a new breed of engineers must be developed who can face the challenges which advanced manufacturing technologies offer. Just as the manufacturing systems integration brings together what hitherto were disparate elements of manufacturing, so the first task of manufacturing engineers is how themselves to bring together the expertise and skills which come from previously disparate technical disciplines.

This process of bringing together implies that the engineer must be skilled at designing whole systems rather than just individual parts of systems. The 'Manufacturing Engineer' must become a 'Manufacturing Systems Engineer' (MSE) whose key skill is likely to be that of the impresario; a coordinator with sufficient organisational computing and engineering skills to design and organise all aspects of manufacturing. This person is not an Engineer in the classical sense of a Mechanical or Electrical

Engineer, but by the nature of manufacturing he or she must be broad based with an understanding of the manufacturing disciplines which affect manufacturing. Nor should the MSE be considered a second-rate engineer. Perhaps the MSE's knowledge in any one of the other engineering disciplines will be less than the specialists, but other qualities will be required that make this broad based engineer a practitioner of equal intellectual standing.

So two related themes emerge. The first is of an engineer with broad based competence in manufacturing technologies. The second is the idea of an engineer who takes a higher level systems approach to the role and contribution of the manufacturing system. In effect this individual falls between the conventional historical disciplines of 'manufacturing engineer' and 'manufacturing manager'. MSEs must master both areas (see Figure 8.1). Some companies, most notably Lucas Industries, have been active in promoting the MSE concept, with remarkable effect.[4]

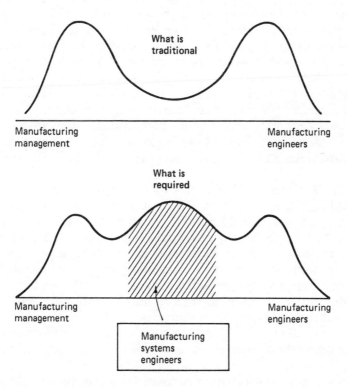

Figure 8.1: Manufacturing systems engineering combines traditional engineering and management skills

The MSE concept is still relatively new but the trend is hopeful. The key development task is to draw connections between the managerial concerns of the manufacturing manager and the 'technical' concerns of the manufacturing engineer. The present gap between the knowledge base of each can become the centre of gravity of the redefined job, and it requires much of the whole basis of engineering education to be similarly redefined.

The first challenge is the theory practice gap in education, to which students seem depressingly resigned. Even in those degree courses with built-in industrial internships the 'academic' and industrial parts of the course are often treated as mutually exclusive learning events. Yet there is enormous potential here for bringing together the best of academic approaches, which rigorously and logically lay bare underlining principles, and the more practical, and immediate requirements of industry. Perhaps engineering education should encourage a more 'clinical' approach of reinforcing experience based practice with theory.

The other significant challenge to manufacturing engineering education concerns the team based, group problem solving activities of MSEs. Manufacturing engineering has always been a more gregarious activity than other branches of engineering. But the trend towards task forces, project teams, and the 'impresario role' is likely to require a far higher degree of group skills. The message for Trainers and Educationalists is that learning through group based tasks will become increasingly relevant. They need help to understand the dynamics of group decision processes, the way to deal with conflict in a constructive manner, and the communication skills necessary for inter- and intra-group problem solving.

Above all Engineers will need to develop a culture of learning which helps them to formalise and capture the learning as it happens and, more importantly, communicate it to their colleagues. This provides the raw materials of what should be an ongoing questioning and debate about the improvement of manufacturing systems. It is participating in this debate which gives our Engineers the knowledge, skills, authority and, above all, enthusiasm to contribute to the coming generation of manufacturing breakthroughs.

Changing methodologies need new skills

Look at the developments in manufacturing methodology (the procedural rather than technical aspects of manufacture) and a number of things stand out. First is the very legitimacy of methodology development as a viable (and usually cheaper) alternative to spending money on technology. Neither an aversion to technology nor an uncritical faith in a technological salvation can get the best out of manufacturing operations. Blending methodology and technology is what is required, but it needs imagination. Methodologies are less easy to define, more fuzzy, than technologies. Bringing them into an operation requires some interpretation.

Take, for example, the most significant methodological shift across the whole of manufacturing, the increased adoption of Japanese influenced manufacturing philosophies. This loose collection of ideas which we call Just-in-Time (JIT) has profoundly changed the way we think of the manufacturing task. Many of the ideas discussed in this book owe a lot to JIT. For example, flexible production leads to fast changeovers and short runs, dependable production leads to increased predictability and lower costs, fast throughput means lean efficient operations, and so on. New (or newish) ideas, developed in a different culture, a different economy and probably a different industry, yet carrying general messages for any and all manufacturing managers. The skill (as highlighted in Chapter 1) lies in seeing behind the bundle of techniques and concepts to expose the underlying messages, then adapting and refining them until the ideas have relevance for ones own set of circumstances. The manufacturing manager or engineer becomes analyst, designer and facilitator. No mean task.

DEVELOPMENT AND OPERATION – THE TWO MANUFACTURING TASKS

Changes in the way strategy is put together, micro processor driven changes in technology and new philosophies of manufacturing all lay increasing emphasis on the design of the operation as opposed to its day-to-day management. So, some argue, why not recognise it by separating the development task from the routine management of the operation. Two types of job and two sets of objectives, so why not two sets of people?

The first set of people – the 'developers' – have a planning and shaping role. They are the ones who are charged with building up the company's manufacturing operation to form an effective competitive force. They need to look forward to the way markets are likely to be moving, they need to look broadly across the whole of the operation to judge the best way to develop each part, and they need to keep an eye on competitor behaviour so as not to be outmanoeuvred. All are tasks which need close liaison with Marketing Planners, Product Development and the company's Accountants. All are tasks which need some organisational 'space' to be performed effectively. Certainly not tasks which coexist readily with the hectic and immediate concerns of running an operation.

The second set of people – those who run the day-to-day operation – have a very different role. Theirs is partly a reactive role, one which involves finding ways round unexpected problems: reallocating labour, rerouting production flow, solving quality problems, and so on. They need to look ahead only enough to make sure that resources are available to meet production targets. Theirs is the necessary routine of manufacturing. Knowing where the operation is heading, keeping it on budget and pulling it back on course when the unexpected occurs. No less valuable a task than the developer's but very different.

But separating 'development' from 'operations' formally in the organisation is not without its dangers. Applied in a naive way it can cause more problems than it solves. It may allow each to concentrate on their different jobs, but it also can keep apart the two sets of people who have most to gain by working together. Here is the paradox, the development function does need freedom from the immediate pressures of day-to-day management but it is crucial that it understands the exact nature of these pressures. What makes this operation distinctive? Where do the problems occur? What improvements would make most difference to the performance of the operation? Questions only answered by living with the operation, not cloistered away from it.

Similarly, the day-to-day operations manager has to interpret the workings of the operation, collect data, explain constraints, and educate developers. Without the trust and cooperation of each, neither set of managers can be effective. Separating them into two 'camps' is not necessarily the best way to promote either trust or cooperation. Most importantly, Operations will have to take a leading part in the implementation of whatever changes are devised by Development – a process which may not be made easier by organisational separation.

The other main problem with separating Development from Operations is the issue of responsibility. Who bears the ultimate responsibility for making sure that the operation is developed in such a way as to improve current and future performance? One danger is that, by default, those who change the operation take on responsibility for the changes rather than the people who will have to live with them in the long run. The worse danger is that no one takes full responsibility, and any initiative falls into the gap between the two function.

Guarding against these dangers is partly a matter of how the players see their roles. Manufacturing Development should neither be the sole originator nor the sole interpreter of manufacturing strategy. Nor is Operations there merely to be told what is best for it. Operations is the function which provides the company's competitive ammunition. So it is Operations, and no one else, who should take ultimate responsibility for developing itself. Manufacturing Development should provide a service to Manufacturing Operations, not try to be its master. In fact it is far better the other way round. The chief client for development services is the Operations function (see Figure 8.2). Theirs is the responsibility so theirs is the right to call the tune.

Most Manufacturing Managers should be general managers

With much of the legwork for development devolved to Manufacturing Development, the Manufacturing Operations function is left with a more clearly focused role – to meet its competitive performance objectives; the key word here being 'focused'. The organisational design must allow each part of the operation to focus on its own defined and limited set of objectives.

The previous chapter pointed out that process technology can be segmented within the plant so that it is appropriate for specific

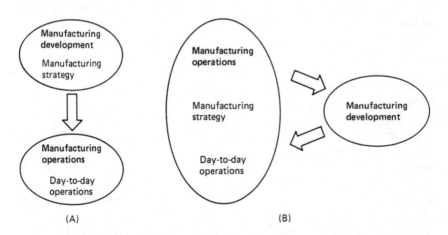

Figure 8.2: Development should not dictate manufacturing strategy to operations (A), it should act to help manufacturing operations link its strategy to its day-to-day operations (B)

sets of product and customer demands. The 'plant within a plant' concept. But the idea of focus goes beyond developing appropriate technology, important though that is. It extends also to agreeing a clear, explicit and narrow set of operational objectives, and developing an appropriate set of skills within each part of the operation. This can be even more powerful an argument for segmented manufacturing than is focused technology. Tightly focused objectives mean being able to say exactly what is and (even more important) what is not expected of each bit of the operation, so organisational ambiguity is reduced. With focused objectives comes the possibility of skills and organisational learning being similarly focused.

It also places a greater emphasis on each part of the operation being self-supporting – having control of most of its indirect tasks. What use are clearer objectives if the segmented parts of the operation do not have control over the resources which are necessary to meet the objectives? This must mean some move towards breaking up the indirect facets of the traditional manufacturing organisation, such as maintenance engineering, manufacturing systems engineering and production control, and decentralising them through the operation.

Doing this has two important implications. First, it makes the direct production chain of command the 'spine' of the organisation. This helps to put organisational form to the objective of connecting operational with strategic management. Second, it pushes the 'general manager' role further down the organisational hierarchy. Sometimes called the 'ship' system, it implies that each manager of a segmented part of the plant, like a ship's captain, should take on responsibility for direction and development of his or her equipment and 'crew'. This mini-general manager concept can become a reality only if these indirect tasks are split up and distributed through the organisation.

However, there is a limit to how far this idea can be taken, and the limit is set primarily by utilisation. If three maintenance staff are sufficient to provide a reasonable service to a plant of four segmented parts, moving a maintenance person into each part seems to require an extra person simply to provide an underutilised maintenance service. Under such circumstances are the benefits of focus worth it? Furthermore the maintenance staff themselves, after years of deriving their status partly through separation and specialisation, would no doubt have views on their dispersal.

There are no easy answers to how far to take decentralisation, but consider the following points when deciding how far down the organisation to push focused units.

- Underutilisation of specialist staff assumes that their specialisms must remain pure and intact. The wider the range of skills the easier it is to achieve higher utilisation.
- Multi-skilling can be a very wide concept. For example, why can't a maintenance engineer also schedule production? Unusual maybe but quite possible.
- Upskilling production staff to perform traditional 'indirect' jobs such as maintenance, quality improvement and resetting machines has been used to great effect in many plants.
- Flexible people tend to be more expensive both in payroll costs and in the training and development they need. But evaluate the costs against the (admittedly less tangible) benefits of quality, throughput speed, dependability and flexibility.
- Always relate the exact nature of proposed organisational changes to the competitive objectives of the unit. The key questions are, 'How should the unit be contributing to overall competitiveness in terms of performance objectives?' so, 'What organisational skills best achieve this?'

Box 8.2

Merging the activities and resources of previously autonomous departments can prove difficult but can also bring considerable benefits. Birds Eye Walls' efforts to integrate its Maintenance and Production organisation at its Grimsby plant bear witness to both the problems and the benefits.[5] The company is the world's largest frozen food manufacturer, and part of the Unilever group. Increasing competition prompted the company to investigate new methods of manufacture to maintain its high quality levels but reduce manufacturing costs.

Part of this policy was to break down the long-standing barriers between Maintenance Engineering and Production to form an integrated Manufacturing operation. This involved the physical relocation of maintenance staff to the production areas. Similarly the maintenance and production supervisors were placed in the same office area. The concept of 'line patrolling' was introduced which supplemented the normal preventative and breakdown maintenance. This involved maintenance engineers roaming the production areas actively detecting problems in their early stages and problem solving to avoid a major breakdown occurring in the future. Line patrolling also meant that the engineers were on hand for a quicker response to any member of the

---

production staff, which fostered a greater involvement between the two groups. More difficult but just as vital was the move towards multi-skilling the engineers. Training was central here, and in spite of some trepidation, especially from electrical craftsmen, the enhanced status (and higher pay) associated with multi-skilling overcame most fears.

Initially most of these moves encountered some resistance. Craftsmen, quite rightly, saw their decentralisation as meaning that they would be more readily observed, easily contacted and would be more accountable. Engineering supervisors also initially saw the moves as lowering their status (which it was not intended to do) and eroding the 'Engineering' hierarchy (which it was). But not all problems were attitudinal. Line patrolling for example really did mean that maintenance staff could not just relax when not on conventional maintenance tasks. Also, supervisors found it more difficult to locate their engineers when they were needed to respond to an emergency.

Most problems were overcome partly because the sound sense in the moves was recognised by most groups. Partly also the process was helped by the company's team-working approach, which it calls 'Workstyle'. The growing acceptance of team based problem solving together with individual appraisals and merit pay for craftsmen contributed to a climate in which most problems could be worked out. Certainly the results were impressive. Operational efficiencies which had been static for years improved by some 15 per cent within two years of the start of the initiative.

Development and Operations need bridging

In the previous chapter the idea of overlapping or simultaneous engineering was discussed as a way of linking product and process technology. It is a concept which has wider applicability than product introductions. It can also be used to think about the relationship between development activities (product or process) and the operations function which certainly will have to implement it, and should ideally have commissioned it.

Overlapping upstream and downstream stages of the development-to-operations process, in terms of the early scheduling of downstream resources, will not by itself ensure an effective interface between stages. In addition, communication between stages must be continuous and effective. Not an easy task, but one helped by formally building 'bridges' between the teams responsible for each stage in three ways.[6]

> *Using procedural bridges* – formally institutionalising procedures such as joint planning and scheduling routines, joint project responsibility and joint appraisal systems.

> *Using 'human' bridges* – such as the rotation and secondment of personnel, encouragement of informal contacts, and the creation of liaison roles.

> *Using organisational bridges* – such as a formal organisational structure to manage the project, task forces, and project managers.

What organisational structure, then, is appropriate to encourage bridging between Development and Operations? Almost certainly some form of project based organisation, although this can take several forms depending on what emphasis it puts on either the functional or the project aspects of development.

Any project will, to some extent, have a life of its own. It has a name, one or a group of people who are championing it, a budget and, hopefully, a purpose. At the same time the many and various activities involved in developing the project could need the attention of several functional groups both in the Development and Operations area. Groups of people organised together because they have some common set of tasks to perform, or common skills to contribute. Which of these two ideas, the project or the function, should dominate the way in which the process of development is managed?

Before trying to answer, look at the range of organisational structures from 'pure' functional to 'pure' project forms in Figure 8.3. In a pure functional organisation all staff associated with the project are based unambiguously in their functional groups. There is no project based group at all. They may be working full-time at the project but all liaison is carried out through their functional heads. The project 'exists' because of agreement amongst these function heads. This probably means that the project passes sequentially from one function to the next with little or no overlapping between stages.

The middle three structures are variants on 'matrix type' project management. From what has been called a 'lightweight project manager' structure on the left, through the 'heavyweight project manager' structure, to the 'Task Force' structure on the right.

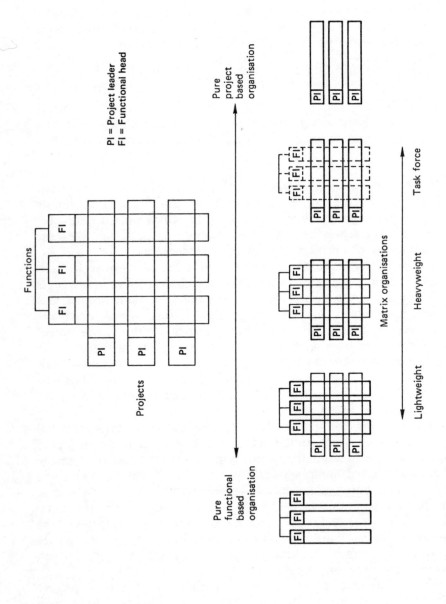

Figure 8.3: The functional project spectrum

In the former all personnel are based primarily in their functional groups but are coordinated by a project manager, possibly supported by a team of liaison personnel, one from each functional area. The direction of the project is in the hands of the project manager who coordinates, liaises and monitors progress. Mainly, though, it is dependent on the functional managers who allocate the resources. Because of this split, the project manager is often a relatively junior person, who nevertheless can have a significant positive influence by being energetic, socially skilled and politically adept.

More project oriented is the 'heavyweight project manager' organisation. Here the project manager has direct control over the activities of the personnel working on the project. They may still retain a base in their function, and they will probably see long-term career development in terms of their functional base, but they are clearly under the direction of the project manager. Because of this the project manager needs to be relatively senior, at least of equal status and authority to the functional heads.

The final matrix organisation involves the specialists from each function being seconded from their functions and physically relocated to a 'task force' dedicated to the project. The task force is led by a 'heavyweight' project manager who probably holds all the budget allocated to the development project. Not all members of the task force necessarily stay in the team throughout the development period, but a substantial core will see the project through from start to finish.

The other 'pure' type of organisation does not have any functional element at all. Staff belong to a task-based group and all authority, communication and resources are channelled through the project team. An unusual form and one which presents some considerable problems, not the least of which is coordination. This is probably why the middle three matrix type structures are the most commonly attempted solutions to managing the Development–Operations interface. What evidence there is seems to indicate that either 'heavyweight project manager' or 'task force' structures are the most effective.[8] They are capable both of consuming fewer resources and bringing the development project to fruition faster than more functional based organisation structures.

Box 8.3

Matrix type organisation structures can become hideously complicated for large multinationals. They not only have the various functions, and business development elements, they also have geography to reconcile. Take for example Dow Chemicals, one of the pioneers of matrix management in the 1960s. They have been almost continually refining their matrix ideas since they found that what seemed an ideal organisational structure in practice generated miles of red tape, scores of committees and a labyrinthine bureaucracy.[7] Multiple reporting channels led to confusion and conflict. Overlapping responsibilities resulted in battles over organisational ground and ambiguity in decision making.

Dow blames its original matrix structure for several poor decisions. In bulk chemicals for example, the economies of scale should dictate a small number of large plants. Yet too many plants were built, largely because of each geographic area's preference for its 'own plant'.

Rather than dictating the form of its matrix structure Dow decided to adapt it – to make it flexible enough to handle different priorities within a single management system. To do this a small team at headquarters sets the business priorities for each type of business. Only after this is the decision taken over which of the three elements of the matrix – product, function, or geographical area – will carry most weight in decision making. This varies according to the type of decision and the type of market.

Dow credits its flexible matrix structure with piloting through a number of successful innovations. For example making its European division the first to buy its oil-derived feedstocks on the spot market, which most chemical companies now do. It also redesigned some of its plants to change over from naptha to liquefied gas feedstocks inside 24 hours. Competitors can take several days to change.

PRACTICAL PRESCRIPTIONS

- Think through the way skills requirements are changing by assessing the changes which have been happening in Strategy, Technology and Methodology.
- Assess how Manufacturing managers contribute to strategy formulation. Identify development needs by identifying failures in how manufacturing is represented.
- If there is a possibility of the operation becoming more accessible to customers and suppliers, decide how operations managers can best interact with them.
- Critically examine the qualities of the engineers who bear responsibility for the well-being of the operation. Do they have the skills to coordinate the parties who have an interest in its development?
- Separate some of the manufacturing development task from the day-to-day running of the operation, but make sure that the latter really is the 'client' in all development activity.
- Make sure that manufacturing managers have as many of the resources as they need to fulfil their objectives within their organisational responsibilty.
- Identify how job flexibility could reduce the under utilisation of resources which might otherwise be involved in decentralising indirect jobs.
- Build bridges between Development and Operations through procedures which encourage joint activity, secondment and informal contacts and formal organisational structures.
- Consider the spectrum of project based organisation structures. The more emphasis given to the 'project' element, the more effective will be the Development (as opposed to the Operations) activity.

9

MANAGING SUPPLY NETWORKS

No operation, or part of an operation, exists in isolation. Every bit is part of a larger and interconnected network of operations. Materials, parts, assemblies, information, ideas and money flow through customer/supplier links from raw material processors, minor and major component suppliers, final product and original equipment manufacturers, distribution systems, after-market operators, through to end customers. Each operation is just one link in a complex network.

That is why a supply network perspective is so important. Only when an individual operation understands the needs and constraints of the other players in its total supply network can it shift its own performance towards becoming a useful and profitable part of the network. Its customers, customers' customers, suppliers and suppliers' suppliers who form the rest of the network, are also partners in the ultimate task of supplying the final customer. If any part of the network performs below par, it reduces the effectiveness of the rest. And like damaged tissue in some living organism, if one part of the network is unhealthy, it needs to be cured, bypassed or replaced if the network is to survive.

Supply networks can be viewed at three levels. Stand far enough back and any operation is a small part of a *total network* like that shown for example in Figure 9.1. But within the total network, and of more immediate concern to the operation, is its *immediate network* of customer/supplier relationships, where the operation is both a supplier to some and a customer to others. Finally within the operation itself is an *internal network* – flows of materials and information between departments, cells or sections of the operation. This chapter deals with supply networks at each of these three levels.

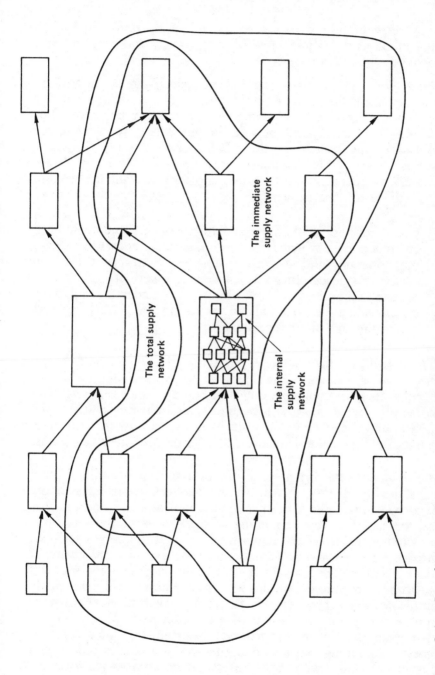

The total supply network

The immediate supply network

The internal supply network

Figure 9.1: Supply networks can be considered at three levels: the total supply network, the immediate supply network and the internal supply network

161

THE TOTAL SUPPLY NETWORK

It may seem grandiose to set an individual operation in such a wide context as the whole network of customers and suppliers, but it does have its advantages.[1]

It puts the operation in its competitive context Immediate customers and immediate suppliers, quite understandably, are of immediate concern to any competitively minded company. But sometimes we need to look beyond the immediate. We need to see ourselves in the context of the whole network. Only then is it possible to see why customers and suppliers act as they do. Because, any operation has only two choices if it wants to be close to its end customers. Either,

- It can rely on all the intermediate customers and customers' customers etc., which form the links in the network between the company and its end customers, to transmit the end customer's needs efficiently back up the network; or
- It can take some responsibility for understanding how customer/supplier relationships transmit competitive requirements through the network.

There really is no choice. Opting for the first is putting too much faith in someone else's judgment for something which is central to its own competitive health. Supply chains, after all, can become obsolete no matter how good some individual parts of the network may be. Companies which are successful in their own terms can still go under because of the failures of their customers.

It helps identify the key players The key to understanding supply chains lies in identifying who in the network contributes to the things which the end customer values. So, as ever, any analysis must start with an understanding of the elements of competitiveness – but this time at the downstream end of the network. Next, the parts of the network which contribute most to end customer service need to be identified. All links in the network will contribute something, but not all contributions will be equally significant. Each part of the network may understand what's important, but not everyone is in a position to help.

For example, the important end customers for some types of domestic plumbing parts and appliances are the installers and service companies who deal directly with domestic consumers. They are supplied by stockholders whose competitive stance relies

---

Box 9.1

The move into mail order selling in the fast maturing personal computer market was prompted originally by the need to trim costs. With most manufacturers (with the important exception of IBM) buying their components from the same group of suppliers, the potential for cost cutting was limited. The growing number of sophisticated second- or third-time customers no longer needed the same degree of technical support from dealers. Cutting them out seemed a good move to Dell Computers, who during the late 1980s became the most successful to bypass the downstream supply chain and deal direct with its ultimate customers. [2]

But Dell found that reshaping its supply chain yielded more than lower costs. Its now direct contact with customers meant that it could learn more about their needs and preferences ahead of rival manufacturers. Realising that its new found potential needed to be exploited, Dell built systems which could track every contact with customers, from their first enquiry through to building a service history for their machines. As well as helping to sell and service computers more effectively in the short term it also let sales and support staff pass better information back to product development teams.

---

on a combination of price, range and above all a high availability of supply – having the part in stock and delivering it fast. Suppliers of parts to the stockholders can best contribute to their customers' competitiveness partly by offering a short delivery lead-time but mainly through absolute delivery dependability.

The key players here are the stockholders. Without keen service and competitive prices from them, the effective end customers will not be as likely to buy your products. It is the end customer who has the real money, and the best way of winning that money is by helping the key players in the network. In this case by giving the prompt delivery which helps them keep costs down while providing high availability.

It shifts emphasis from short-term opportunism to long-term profitability There are times when the tides of organisational fortune render parts of a supply network weaker than its adjacent links. How then should its immediate customers and suppliers react? Exploit the weakness as a legitimate move to enhance their own position maybe? Or ignore the opportunity, tolerate the problems, and hope the customer/supplier will eventually get back on its feet?

Perhaps neither option is in the best long-term interests of the whole supply network. Yet both options are commonly pursued.

Short-term adversarial opportunities seem too good to miss, and short-term issues too pressing to give thought to how the total supply network is being affected. A longer-term view would be to weigh the relative advantages to be gained by the whole chain from assisting or replacing the weak link.

While a routinely established habit of considering all external relationships as part of the total supply network is not a guarantee of long-term self interest or an antidote to short-termism, it does help. When viewed in the total network context, it seems less sensible to either ignore or exploit a part of the network on which you depend to serve your own end customer.

It avoids local fixes It is futile to attempt long-term improvement in one link of a supply network without consideration of how the changes will affect players in other parts of the network. It is not the improvement itself which is futile, it is the parochial nature of the improvement and the isolated objectives to which it is addressed.

At its best such localisation merely fails to have any effect beyond adjacent links up or down the network. At its worst it shifts problems elsewhere in the system.

It sensitises operations to macro changes Supply chains in many industries are undergoing fundamental structural change. Specialisation and focused plants means more fragmented supply chains, each link of which buys in a greater proportion of its costs. Global sourcing and changes in the relationships within and between trade blocks introduce (maybe) temporary turbulence into supply relationships. For example the consolidation of the supplier base as just-in-time ideas establish themselves in some industries. The supply network is rarely a totally stable context in which to do business. Again, a total supply network perspective does not guarantee that a company will fully understand all significant changes in its trading environment. Nor does it necessarily mean that the company will be able to devise strategies to cope with the changes. But it does give it a fighting chance.

Box 9.2

Many supply networks split into two or more chains after a point. Take the supply network for motor vehicle components for example. It splits after the component manufacturers to supply the motor vehicle manufacturers in one chain, and the spares 'aftermarket' for components in another.[3] See Figure 9.2.

In fact this particular network demonstrates some typical characteristics of supply networks.

• Different end customers have different needs, even for the same products.

The spares market is largely a 'distress purchase' market, with priority placed on parts availability or responsiveness rather than price. Motor vehicle manu-

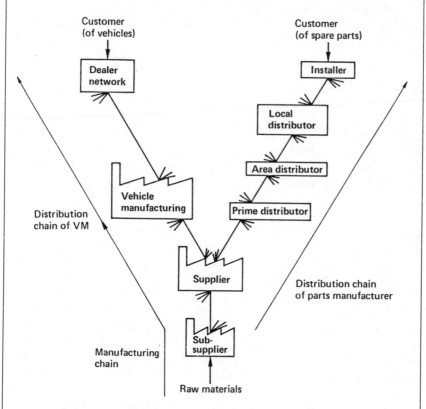

Figure 9.2: *The automotive components supply network splits after the component manufacturer*

facturers, on the other hand, are very much interested in quality and price, together with JIT delivery.

- The actual end customer is not always the most important end customer.

  The actual end customer (the consumer) in the spares leg of the supply network does not always differentiate between different products. The important end customer is the last person who makes a decision regarding which actual product is bought. In this network it is the Installer.

- Some parts of the downstream network stand out as current, or potential key players.

  In the automotive aftermarket network the specialist and 'menu' installers are seen as being especially important end users because of their potential for standardisation, branding and service control. It makes sense therefore to develop supply network performance to be in line with these important downstream players.

How much of the chain should you own? – Vertical Integration

Vertical integration is the extent to which a company owns the supply network of which it is a part. In its strategic sense it means assessing the wisdom of companies acquiring suppliers (upstream vertical integration) or customers (downstream vertical integration). At the individual product level it concerns the make or buy decision.

Notwithstanding their fascination with the topic it is seemingly impossible to get a straight answer from economists to a straight question, 'Is vertical integration good or bad for a company?', or, if 'it depends', what does it depend on? But they cannot entirely be blamed for their ambivalence; the decision is complex.

Yet the rule of thumb is relatively simple, even if the decision is not. Do the advantages which vertical integration give in your particular set of circumstances match the operational requirements for you to compete more effectively in your particular market? Don't rely on too broad generalisations. Rather let the decision be tailored by competitive priorities. So for example, if external performance objectives for the operation are dependable delivery and meeting short-term changes in customers' delivery requirements, the key question is, 'How does vertical integration enhance dependability and delivery flexibility?'

This means balancing two sets of opposing factors: those which give the potential to improve performance, and those which work against this potential being realised.

As far as product quality is concerned, the potential of vertical integration comes from closeness to customers and suppliers. The sources of quality problems can be easier to trace through in-house operations and the subsequent quality problem solving can be concentrated at the most appropriate point in the chain. Acting against this is the danger that in-house operations, freed from the disciplines of a true commercial relationship will have less incentive to cooperate.

As far as delivery speed is concerned, vertical integration can mean a closer synchronisation of schedules which speeds up throughput along the chain. Also, being close to both supply and demand helps better forecasting both of customer needs and supply constraints, which reduces the risks of speculative manufacture. Again these potential advantages can be eroded if assured 'in-house' business means that in-house customers get low priority compared with 'proper' customers.

As far as delivery dependability is concerned, improved communications along the vertical integration chain can mean better forecasts and more realistic delivery promises. Even when internal hold-ups mean that deliveries will be missed, in-house suppliers are more likely to give notice of the problem and therefore better delivery integrity. All of which again assumes that the relationship between vertically integrated links will indeed receive top priority rather than being overlooked – the victims of having little conventional commercial power.

As far as new product flexibility is concerned, vertical integration gives the potential to guide technological developments as well as deny them to competitors. Downstream vertical integration gives the potential for products to be developed specifically and more precisely to customer needs. The danger is that if management attention is spread too thinly along a vertically integrated chain, opportunities to exploit structural links are missed through dissipation of attention.

As far as volume and delivery flexibility are concerned, ownership of suppliers can give the potential to dictate volume changes to match downstream fluctuations as well as helping to expedite specific orders through the supply chain. Against this there can be a reluctance to inflict volume changes on in-house suppliers and customers. It is perhaps easier to be commercially realistic with independent companies. Further, even with common ownership the various technologies of the supply chain can have widely different break even points and capacity increments. Ownership does not change the basic economics of capacity management.

As far as cost is concerned, shared ownership can provide the potential for shared overheads – R&D and logistics, for example. And over the longer term can allow capacities, and therefore capacity utilisation, to be balanced. Perhaps more significantly, if margins are high in supplier operations, why not, 'capture the margins' and reduce the costs of bought-in parts. Of course this assumes that other customers of a newly acquired supplier will be content to continue doing business with it. If they are not, volumes could fall and unit costs increase. Even if volumes sustain, will a management now concerned with more separate businesses be as efficient managers or make investment commitments away from their core business?

Does vertical integration really matter?

It does not need too close an examination of the arguments for and against vertical integration to see that much of its potential lies in the assumption of closer communication and cooperation. Conversely the main factor working against the potential benefits being realised is the tendency of companies to take their in-house operations less seriously as commercial customers and suppliers.

Both points prompt two questions: whether the closeness sought through vertical integration can be gained through other forms of relationship, and whether the familiarity which leads to contempt for in-house customers can likewise be avoided? Many companies think that the answer to both questions is a qualified yes. Partnership arrangements with suppliers and customers can be constructed which lead to cooperation and shared interests while maintaining the commercial bite of a relationship which exists only because it is in the long-term interests of both parties. Such 'partnership' arrangements are part of developments in how companies are managing their 'immediate supply network', discussed next.

THE IMMEDIATE SUPPLY NETWORK

Of all the customer-supplier links in a supply network, the important ones for most companies are those with their own immediate suppliers and customers. It is no use developing perfect understanding of the total network if immediate links are

168

neglected. In fact the point of knowing how the network operates is to manage the immediate links more effectively.

Analysing customer-supplier relationships.[4]

For most industries the nature of the supply relationship has changed fundamentally over the last decade. This is true especially in markets subject to high levels of competition. In automobile and consumer electronics industries, for example, traditional customer-supplier relationships based on an arm's length, price based involvement, broke down during periods of stress brought about by competition and recession. What has emerged is the idea of a 'partnership' with suppliers – an ideal supposedly espoused by all progressive companies.

The 'partnership' model sees customer-supplier relationships as being based on openness, trust, shared destiny and long-term development. Partnership means a more exclusive relationship (fewer suppliers, simpler networks) and a 'richer' relationship in the sense that more than orders and parts flow between operations. So do information and longer-term plans. Product development responsibility can shift to suppliers, market forecasting responsibility to customers, and so on.

Box 9.3
Developing suppliers towards the partnership model depends, in part at least, on how they are introduced to the responsibilities and opportunities of their new role. One study[5] suggests that the most effective moves towards achieving the new relationship start with face-to-face discussions between the chief executives of customer and supplier. With larger customers, although high-level management contact is important, so is the way customers communicate their supply philosophy to vendors. Ford Motor Company produced a video on their quality requirements, for example. Nissan Motor Manufacturing (UK) hold seminars for potential suppliers where senior management outline Nissan philosophy. IBM hold road shows where IBM representatives visit suppliers with examples of products so that all staff can see where their components fit in IBM's final product.

Yet is the new model relationship quite as cosy as it is portrayed? Should we all now be close and trusting allies of our suppliers? Not quite. The danger is believing that with openness and honesty comes a relaxed easy relationship.

The Lean Supply Model

Richard Lamming, who contributed to a major world-wide study of the automotive industries, developed what he called the 'lean supply model' – a picture of the way customer–supplier relationships in the industry are moving. But it has lessons which stretch beyond the automotive industry.

Competition will continue the trend towards fewer, more technologically sophisticated suppliers who collaborate more closely with customers. *Sourcing decisions*, while not ignoring price, will be made largely on the service history demonstrated by the supplier and the customer's view of the supplier's capability to join it in the development of new products. This means an earlier involvement of predetermined suppliers for the development and supply of each individual component.

The alliance between customer and supplier will require a far greater transparency of *information* on intentions, future plans, designs, and more than anything, costs. This in many ways is the basis of the whole relationship. Suppliers relying on their capability and track record to win business, and not attempting to withhold information from customers to achieve short-term bargaining advantage. Openness could even extend to exchange of (usually technical) personnel for development and improvement.

If all this sounds too ideal, consider that these changes in the nature of the relationship are not pursued for their own sake. They are strictly a means to a clearly defined end – more profitable operations based on the sheer excellence of service which is in the mutual interests of both companies. Further, this is not achieved without considerable stress in the supplier organisations. Pressure will come certainly from the customer, but also will be generated within the supplying company itself. 'In lean supply, the pressure is self-imposed. The lean supplier must actually drive himself harder than the customer does. Only in this way will the supplier be able to take the initiative in R&D, total quality control, and so on.'[6]

The quest for superiority inherent in the lean supply relationship is reflected in how it manages its performance objectives.

Quality management aims beyond supplier quality assurance schemes and towards the state where suppliers do not have to be pushed into improving quality. They will already have been encouraged to develop their processes to a point where they are themselves leading customers through a programme of continual interaction which sets mutually agreed targets.

Delivery speed, in terms of customer lead times, continues to improve largely through JIT efforts speeding up materials and information throughput times. More significantly customers should take on more responsibility for clearly articulating their lead time requirements rather than expecting suppliers to respond to unexpected requests.

Dependability becomes even more vital as customers delivery times are specified to the hour. Production schedules, synchronised between customers and suppliers, help to achieve high dependability but the move towards global sourcing of some parts could mean managing differences between local and long-distance suppliers.

Flexibility, in terms of new product flexibility, becomes a combination of two modes. Partly 'black box' development led by the supplier, where the customer, having entrusted development to the supplier, is more concerned with the way the component fits into the product rather than its internal details. Partly it will mean integrated activities where longer term development is achieved through integrating expertise. Volume flexibility, at least in the medium term could mean closer matching between supplier and customer's output levels as neither are willing to bear the costs of the inventories which would otherwise accumulate between them.

Cost and Price relationships must undergo an even more profound change. Traditionally a common automotive industry tactic was for the supplier to quote an initially low price, perhaps even below its cost, in order to secure the order. It could do this, secure in the belief that there would be plenty of opportunities to raise prices in later years. So it could more than recoup its initial sacrifice (see Figure 9.3). Lean supply however requires habitual cost reductions (and hence price reductions) based on total transparency of cost information and an unshakeable belief that the more a product is made, the lower should be its manufactured cost. To the supplier, cost reduction through continuous improvement becomes a way of life governed by the discipline of the annual expectation of price reductions; but with some security coming from the knowledge that, knowing your cost structures, it is in your customer's best interests to make sure that you are

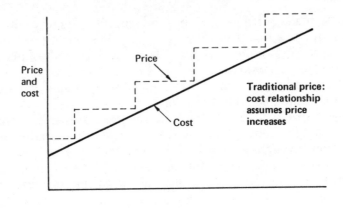

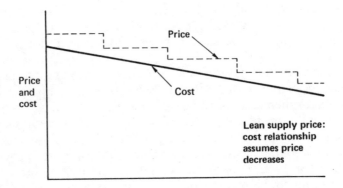

Figure 9.3: Lean supply expects cost to reduce and supply price to reduce with it

making a reasonable profit. Some security perhaps, but not enough to make life comfortable. Hence the high stress in the relationship.

Overcoming the Problems of Lean Supply

Suppliers can get understandably anxious that the hidden (or not so hidden) intent of lean supply is merely to transfer the problems and neuroses of customers back up the supply network. Sometimes such anxieties are well founded. Take one example.[7]

Relationships between a supplier and a customer in the automotive industry had deteriorated over a number of years in spite of them doing business together for many years before that. The customer's discontent stemmed originally from a decision by the supplier some three years previously to increase its prices substantially to what it considered a 'market rate'. Furthermore the supplier decided to charge different (higher) prices for parts sold to the customer's service (spares) organisation than it did for those sold to the plant. To compound the customer's grievances, the supplier insisted on a 30 day payment period when the customer requested 60 days from all other suppliers. And the supplier would maintain only a five-day buffer stock whereas the customer felt it needed a ten-day buffer.

On the other hand the supplier regarded the customer as its most difficult because of frequent schedule changes which caused considerable problems. For example, some items in the supplier's inventories once manufactured could be in stock for several months.

The customer's excuse for short-notice schedule changes was the volatility of its market. In reality market volatility was considerably compounded by the customer's planning system which took several weeks to convert sales forecasts to material orders. The supplier, on the other hand, operated a pull scheduling system based on some features of JIT manufacturing. However, although in some ways the system was responsive (small changes in mix could be accommodated at short notice) factory capacity targets were set several months in advance. Major aggregated schedule changes were resisted by the supplier. The two manufacturing planning and control systems were incompatible both in their philosophy and in their operating details

The position was further complicated when the customer, partly as a punitive measure, decided to second source some of the business previously placed with the supplier. It eventually selected a Japanese supplier who was price competitive but demanded (and obtained) stability of customer orders. This new supplier took around 30 per cent of the original supplier's business. With the stable element of orders taken from the supplier, volume and mix fluctuations became even more erratic. The mutual dissatisfaction increased and commercial negotiations between the two companies reached stalemate.

The two companies did eventually pull back from the brink of what could have been the total collapse of its trading relationship. Dual sourcing was continued but not for individual part numbers, so variety was reduced and, because the fixed element was no longer taken from the schedules, so was the variability of orders. Both manufacturing planning and control systems were modified. The customer's system was speeded up and forecast data extracted at an earlier stage. The supplier also agreed to make its master scheduling system as flexible as its pull based manufacturing system. Negotiations over the commercial relationship were restarted and some compromises over payment periods and stock holding (made easier by the moves towards planning and control system compatibility) reached. Finally electronic data interchange (EDI) was introduced to speed up exchange of data.

A number of lessons come out of this example which have general applicability.

First, commercial issues are critical. There is no point in the Purchasing or Quality departments working on the relationship if Credit Control are not paying the bills for 60 days. Commercial details set the climate in which material and information flow improvement can be made.

Second, manufacturing planning and control systems at both the supplier and customer end must be mutually understood and made compatible. In particular it is helpful if customer's long-term forecasts are made available to suppliers.

Third, demand variability needs to be addressed squarely. Identify the causes of variability and attempt to reduce it as far as possible. At the very least distinguish between the 'fixed' and 'adjustable' periods of the order cycle.

Finally, speeding up the transfer of data holds considerable potential, but only if the other issues are addressed. EDI on its own yields little benefit.

The Supplier/Customer Power Balance

The unstated assumption in much of the discussion thus far is that suppliers are sitting there waiting to be developed. All they need is to be told how important they are, and what an important role they can play, and they immediately accept advice from their benign customers, change their ways and become model suppliers. But for many companies, suppliers are not so easily persuaded. Not because of any obvious incompetence (although long-term short sightedness maybe) but because the customer's business accounts only for a small proportion of their total output.

When a company's position in the supply market is weak compared to its suppliers it finds itself with far fewer options. It certainly should not ignore the approaches to supplier development which are open to more dominant companies, but it will have to accept that persuasion will be tougher and other options need to be considered.

The first option is to explore the possibility of avoiding having to rely on supply at all by bringing manufacture in-house. Alternatively, other components or some limited design rationalisation could be considered. Even if taking supply from a stronger player in the network cannot be avoided there are still options. These include

- concentrating as much volume of total orders on fewer suppliers to gain as much leverage with what buying power the company does have,
- grouping together with other customers to increase bargaining power, and
- negotiating long-term contracts both to guarantee supply and to increase the perceived value of the business to the supplier.

Most important, though, is recognising that as a relatively weak player, extra effort and creativity is needed in seeking ways to overcome supply vulnerability.

THE INTERNAL SUPPLY NETWORK

The internal supply network is in many ways a microcosm of the external network. Each department is both customer and supplier to other parts of the plant. It is an internal flow system which can be managed, in some ways, very much like the external network. But internal networks have the advantage that the internal network is (or should be) better understood, and the ability to influence it directly is greater.

In practice, though, it can seem just as troublesome to manage as the external network. Partly this is because of the detailed complexity with which Manufacturing Planning and Control (MPC) systems have to cope. Partly it comes from a confusion over the basic philosophy of managing internal networks. And partly it is because all the problems of dealing with the external network meet at the factory itself. Irregular, uncertain or badly coordinated supply, and constantly changing demand make the MPC task particularly difficult.

Push and Pull philosophies

The debate between either emphasising the coordination benefits from vertical integration, or looking to each part of the network to develop advanced supplier/customer relationships, has its internal parallel. The alternative approaches to managing the internal network are usually seen as either of the following:

- Coordinating every part of plant's manufacturing by taking a view of what production is required, and from detailed knowledge of product structures, process times and product routings, calculating exactly what each part of the plant should be making and where they should send it to be processed further. This approach is usually called a 'push' approach because instructions from the central MPC system push batches of parts or products through to the next link in the internal network.

- Allowing each part of the plant to act as a quasi-customer to the other parts of the plant which supply it. Rather than have a central MPC system which coordinates every part of the plant, let the instruction to process a batch come directly from the department which wants it next. The 'instruction' can be a simple internal order (the 'kanban') which triggers supply. Each department 'pulls' batches of goods through the system by making sure that goods are processed only when they are needed. Provided that the whole internal network obeys the simple rule of producing only when instructed by a kanban from its internal customer, goods will move quickly through the system, spending the minimum of time as inventory.

This debate is usually represented crudely as Materials Requirements Planning (MRP) versus Just-in-Time (JIT). In fact it isn't quite as simple as that. It is not a matter of choosing between two 'solutions', like choosing the better of two washing powders. The two 'pure' systems can be blended to suit an operation's individual competitive needs. And while it is not in the scope of this book to describe the details of MRP or JIT, we do need to identify what have come to be seen as the advantages and disadvantages of each before deciding how to blend the two.

Pushing on through to the customer – MRP

In the early 1970s when MRP first began to impact on production management, it was greeted with enthusiasm for three good reasons. First, it was such patent common sense compared to the rather academic approaches hitherto. Most demand on the upstream parts of the internal supply network was clearly derived from downstream demand. Why ignore this? Why not derive parts manufacturing schedules directly from finished goods schedules?

Second, it offered a straightforward set of prescriptive steps – it tells you what to do and when to do it. Third, it used the power of computer technology — then starting to become cheap enough to seem a waste not to exploit it.

The paradox of MRP is that it tries to be literally a just-in-time system.[8] Its objectives are to make sure that the plant produces goods just as they are needed by the market. MRP starts by looking forward and asking what end items need to be shipped at what time *in the future*. This is an important point; MRP can plan production when we want to anticipate future output requirements. It uses the bill of materials to calculate how many of which parts need to be ordered from the upstream parts of the internal supply network and from this, how many parts and materials to order from suppliers. By doing this it connects the demand side external network with the supply side external network. It also allows for the time each stage in the process needs to complete its processing task, by assuming a particular (but fixed) lead-time and placing orders at each stage in the internal network at the appropriate time.

It is this last point which highlights the second, but more problematic, paradox of MRP. In order to calculate exactly when departments in the internal network should start processing batches of product MRP needs to assume a fixed value of processing lead time. But in practice the time which a department takes to produce a batch depends on the loading of work which comes from the MRP system itself. The logic is circular; the system needs to be given a realistic lead time, but the lead time depends on the production schedules produced by the system! The obvious way round the paradox is to assume a long enough lead time to cover most circumstances; then at least the chance of failing to meet internal delivery dates is minimised. But what of fast throughput and the benefits of speed described in Chapter 3? What incentive is there to reduce internal lead-times? Certainly none from the MRP system. That does not inevitably mean long lead-times, just that one cannot look to the MPC system directly for throughput improvement.

This is not the only criticism of MRP based systems. The other main practical gripe is that they are just too cumbersome. That they take complex problems and make them more complex with a thoroughness which only advanced (and expensive) computer technology can manage. That they need perfect data which, when not accurate, makes the system do more harm than good. Listen to the following description of one MRP system in a small company.[9]

Feedback from the shop floor was very poor . . . schedules were very inaccurate and fell into disrepute . . . problems were encountered with the software . . . stores accuracy was very poor . . . there were numerous problems over responsibility . . . After six years of operation – during which the company had been persuaded to buy upgrades to its computer and further software – few benefits had been obtained.

So with such problems why do so many companies persist with MRP type systems? The answer is that there is no better way of planning for anticipated changes in demand in a complex operation which needs materials planning and coordination at a high level. The problems come when the system delves into the complexity of detailed scheduling and control. At the longer term, higher level, end of the planning task any manufacturing operation needs the type of production data base which MRP type systems give. Not only does it establish a discipline on longer term planning it acts as the natural focus of interfunctional debate on production priorities.

Pulling from customer demand – JIT

Again let us use the term JIT here to mean the pull-based system for organising materials flow described earlier. It is important to remember though that the JIT philosophy is far broader than a system for dealing with materials flow.

So does JIT overcome the problems of MRP? Well, it does overcome some of them. It clearly promotes faster throughput which gives all the benefits described in Chapter 3 – faster response, lower inventory, and a leanness of production which encourages improvement in efficiency. Furthermore it is a simple and simplifying concept: make only when your internal customer needs it; take responsibility for your own lead time and inventory; keep production more or less level within one period. All these aspects of JIT control don't amplify complexity, they reduce it.

The consequence of which is that JIT type systems place far less, if any, reliance on computing power to drive them. Control and responsibility for making the system work is left at local line, cell or departmental level. So there is no complicated computer system and few worries over degrading the system with bad data. This certainly means that JIT systems need less expensive support and might explain (but does not excuse) the antagonism of some JIT proponents to computer based systems.

The paradox of JIT is that there are circumstances where it is less capable than MRP of achieving literal 'just in time' delivery. This is because pure JIT is a *reactive* idea – it responds to changes in demand only with difficulty. It is not of itself a system which anticipates demand. That is both its virtue and its limitation. The very system which protects upstream parts of the internal network from fluctuations in demand makes it less capable of reacting flexibly to the changes, especially where complexity of the product structure is already straining JIT's need for simplicity.

Box 9.4

In their review of how JIT is being implemented Chris Voss and David Clutterbuck[10] highlight, and knock down, a number of myths surrounding it. They form a list of the excuses which companies use for not considering JIT principles.

- MRP is an adequate substitute for JIT. More appropriate for Western manufacturing. No – first, the two ideas are not exclusive; second, MRP is far more easily installed when JIT ideas are already in place.
- JIT only works in high-volume batch manufacturing such as automotive or electronics. No – there are plenty of examples in, for example, pharmaceuticals, paint manufacture, whisky bottling and distribution.
- JIT won't work in low volume manufacturing. No – the principle still holds and may even be easier because batches will already be small. For example, Hewlett-Packard's computer systems division, which builds less than six computers a day, has obtained considerable benefit from JIT.
- JIT requires daily delivery from suppliers. No – only when its internal network is operating in a JIT manner will an operation benefit from JIT supply. By then it will be in a better position to help its suppliers.
- JIT will cure all production problems. No – many production problems are outside JIT's scope, and over-inflated expectations are a sure way to undermine realistic JIT projects.

Blending JIT and MRP – the Hybrid Systems

The strengths and weaknesses of the two approaches to managing the internal network give the best guide to how and where they should be used. Figure 9.4 takes two of the dimensions which determine the suitability of MRP and JIT – the type of the production process and the level for which the control system is being designed.[11]

The *type of production* is really a combination of factors which indicate the complexity of manufacturing. Variation in processing lead-times, the number of alternative product routes, the complexity of product structures, and the variety of product types, all taken together paint a picture of how complex is the production control task. So for example most chemical processing and paper manufacturing would be towards the left side of the horizontal

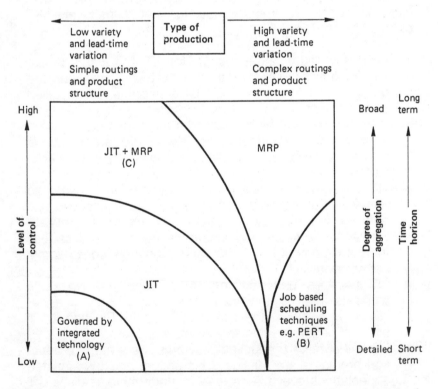

Figure 9.4:   The appropriate planning and control mechanism for the internal supply network will depend on the complexity of manufacture and the level of control

scale in Figure 9.4, consumer electronics and motor vehicle manufacturing further towards the middle, batch engineering manufacture past the mid point, and customised jobbing manufacturing on the right. The scales are not exact, however. Not all factors compressed into this single scale will necessarily go together; some manufacturing might rate as high on some factors but low or medium on others, but this is only a general guide not a complete prescription.

The *level of control* indicates which set of production control tasks are being considered. High-level control involves broadly coordinating the flow of materials to the various parts of the plant as well as giving an indication of what level of output they will be expected to achieve in future periods. Medium-level control is the detailed allocation of production orders to each part of the plant. Low-level control is the detailed monitoring and readjustment of day-to-day shop floor activities. Again the boundaries between the categories are not clear cut but represent graduations in the degree of aggregation and the horizon time of control.

Three of the areas in Figure 9.4 need some explanation.

The area A indicates that in 'process' type manufacture the shop floor level of control may be incorporated into the technology itself. For example, the integrated technologies of some food processing plants automatically transfer materials from one part of the plant to another. It usually requires intervention on the part of production management to prevent transfer occurring.

The area B represents the detailed shop floor scheduling and control in very complex, high variety customised manufacture. Here it is the nature of each individual job which dominates the production control task. Specialised techniques such as network planning (PERT for example) are usually needed.

The area C represents a relatively new set of developments. Several modified MRP approaches are available which add pull type ideas, 'rate based MRP' for example calculates production rates (rather than batch quantities) for items which have some degree of commonality across different end products.

PRACTICAL PRESCRIPTIONS

- Look beyond immediate customers and suppliers to the total supply network.
- Use the analysis of the total supply network to understand suppliers' and customers' behaviour.

- Identify the 'key players' in the supply network, and what they have to do to make the network more effective.
- Address issues of vertical integration by balancing the things which have the potential to improve performance against those which work against the potential being realised. Do this for all performance objectives – quality, speed, dependability, flexibility and costs – giving prominence to the elements which contribute most directly to competitiveness.
- Develop supplier relationships on the lines of the 'partnership' model. But don't expect the relationship to be stress free.
- Always address commercial issues of supply along with operational improvement.
- Make sure that MPC systems are compatible between supplier and customer.
- Develop a shared understanding on how short-term demand variability will be treated.
- Speeding up information alone is rarely helpful. Unless the quality of the information is appropriate one is simple 'speeding up rubbish'.
- Think about the internal supply network in the same way as the external one. Are its capabilities and supply relationships balanced and appropriate?
- Blend the elements of Push and Pull to suit the level of control and the type of manufacturing in terms of volume and variety.

10

FORMULATING MANUFACTURING STRATEGY

The acid test for Manufacturing Managers is not just their understanding and command of the detailed complexity of Manufacturing Operations, important though these are. Rather it is whether they can make enough sense of the operation to fit it into a strategic context, reshape it, improve it, and make sure that its contribution to competitiveness is clear and continuing. Doing this means formulating a set of policies, plans and improvement projects which, when they are taken together, define the direction of manufacturing until it becomes the source of competitive advantage. Devising these policies, plans and projects, is the manufacturing strategy formulation process.

An effective manufacturing strategy clarifies the links between overall competitive strategy and the development of the company's manufacturing resources. It should be able to answer those important 'so what' questions. For example, 'We intend to compete through aggressive pricing – so what does that imply for the way we develop process technology?' . . . 'We have customers who have different requirements for different product groups – so what does that imply for the way in which we set performance targets?' . . . 'We operate in a turbulent market with frequent product changes – so what does that mean for the way we organise the manufacturing function?'

A FORMAL STRATEGY IS WORTH THE EFFORT

So why bother to devise a formal strategy? The effort after all is likely to be considerable and not without its difficulties. Difficulties which can sometimes prove dispiritingly difficult to overcome, and which are especially acute for Manufacturing.

The first difficulty is the dispersal problem. Manufacturing managers are central to the strategy formulation process, yet they, more than most, are likely to be geographically dispersed among the company's sites. As one manufacturing director put it, 'My Marketing colleague has his senior people all within a few steps of his office. My senior people are spread around the country. The effort of getting them together is not something we can go through every week.' Second, manufacturing managers operate in 'real time'. They cannot allow their attention to drift from the running of the plant for anything but a relatively short period. This responsibility for the day-to-day running of the plant means that they operate under an operational imperative from which only the most pressing strategic pressures can divert them. Third, the inertia of operational resources imposes a certain amount of conservatism on whoever manages them. With the majority of the organisation's resources under their control no manufacturing manager will, or can, lightly change operational direction without very good reason. Laudable though such caution might be, it does mean a degree of circumspection which at times mitigates against imaginative strategic change. Finally, the cumulative effect of all these pressures have taken their toll over the years. Manufacturing managers often are just not used to thinking, acting or influencing the organisation in any strategic manner. It is this latter reason which often seems to be behind the inability of manufacturing to represent themselves in the strategic councils of the organisation.

Yet the effort of strategy formulation can be more than just worthwhile. It can be invigorating, a process which provides direction and purpose for what is a complex task. A manufacturing function which knows at least what it is trying to do surely must stand a better chance of succeeding. Certainly, without either direction or purpose, success is reduced to a matter of chance. So if operations managers are too busy to contribute to strategy then no one should be surprised if manufacturing's interests are not represented as strategies evolve.

A formal strategy helps to ensure that the policies adopted in the manufacturing function hang together in a coherent manner. With a common purpose and a set of priorities, conflicts are exposed and debated. For example, a company's technology policy, amongst other things, shapes the integration between different parts of the manufacturing process. Its Manufacturing Planning and Control policy has to work within the constraints imposed by the technology. A shared strategy allows not only both areas to measure their own decisions against the common purpose, but also allows the implications on each other's policy areas to be explored. A formally constructed manufacturing strategy gives the basic structure to ensure that the many individual policies and decisions, taken around the organisation all point in the same direction.

The major benefit of going to the effort of strategy formulation though, is even more far reaching. A credible manufacturing strategy reinforces the centrality of competitiveness in the culture of the organisation. It does this by concentrating on the linkages between overall company strategy, manufacturing objectives, the various manufacturing tasks, and the individual resources of the manufacturing system. An effective manufacturing strategy should bring the concept and feeling of 'competitiveness' right to the shop floor – the very heart of the company.

What should a manufacturing strategy be?

Before describing the process of strategy formulation in detail it is sensible to set down exactly what it is hoped to achieve from an effective manufacturing strategy. Only then can its ultimate effectiveness be judged.

It should be appropriate If the process is to connect operations to some concept of competitiveness then above all it must provide appropriate solutions. In other words the strategy should direct manufacturing change in the direction which on balance is the most likely to provide a manufacturing performance which best supports the company's competitive strategy.

It should be comprehensive A manufacturing strategy cannot pronounce on every minor operational decision but it does have to indicate how each part of the function contributes. No part of the manufacturing function is without influence on performance, therefore no part should be left without guidance. Production, Engineering, Quality Control, Manufacturing Planning and Control, Maintenance, Purchasing and Materials Control must all be included. To fail to include any one of them is to fail to bind them together.

It should be coherent Including all parts of manufacturing in the strategy is a necessary but not sufficient condition for effectiveness. The policies recommended for each part of the function must all point roughly in the same direction. Potential conflicts between the various areas need addressing directly.

It should be consistent over time While no organisation benefits from an overly rigid strategy, the lead time of manufacturing improvement means that consistency must be maintained over a reasonable time period. Failing to provide consistency confuses the organisation. But worse it leads to cynicism . . . 'last year was the year of quality, this year it's short lead-times, what will be in fashion next year?'

It should be credible A strategy which is not regarded as achievable by the company will not be supported. Its subsequent failure will merely reinforce the perceived futility of the whole process. Improvement targets should be seen as feasible to be worth anything.

THE PROCESS

Putting any strategy together is not quite the clinical, systematic affair it is sometimes held to be. Nevertheless there are a number of separate activities which, when put together, can form a logical process. Not that any process should be followed slavishly or the whole activity degenerates into a kind of painting by numbers. But some kind of structure to the process provides something to cling to when the going gets difficult.

There are several ways of putting the stages of strategy formulation together. Most consultancy companies, for example, have their own methods. However, these procedures tend to follow a similar pattern and have a common philosophy known as 'gap' methodology. Put simply (as befits a basically simple idea) gap methodology means four things. First, it means developing a specific idea of what should be important to the manufacturing function in order for it to compete effectively, answering the question of what its objectives should be. Second, it means assessing the actual achieved performance of the manufacturing function. Third, the gap between what is important to the operation and what performance it is achieving drives the priorities for performance improvement. And fourth, the performance priorities govern choice and implementation of long-term and short-term improvement plans. The remainder of this chapter follows these four steps through.

Step 1 – Setting Manufacturing Objectives

The start point for any functional strategy must be to examine its role in improving overall competitiveness. The key questions, which again we return to, are 'How do we want to compete?' and 'Therefore what do we need from our manufacturing function to enable us to compete more effectively?'

The questions may be basic but they are rarely that easy to answer. Certainly the questions cannot be answered without reference to the view of competitiveness taken by the other functions in the company, most particularly Marketing and Product Development. Marketing after all should have a better idea than anyone else of what sells the company's products, how customer's needs are developing and what moves competitors are making. Likewise Product Development should be able to chart how the features, attributes and technology of the product can be developed in the future.

In theory the picture is of a cosy trio – Marketing, Manufacturing and Product Development – each contributing their particular part of the overall picture towards the point where compromises can be struck and consensus reached. Sadly the process is rarely so straightforward, for any number of reasons. The following are typical.

- Different (and sometimes conflicting) views of strategic objectives are held by senior management, e.g. long-term growth in sales volume versus short-term profitability.
- No formal marketing strategy exists.
- Marketing strategy is in a form which is not particularly useful to Manufacturing e.g. market strategies framed in terms of 'prime customers' rather than the competitive stance of each product group.
- No consensus exists on how the market is developing.
- Product development have no long-term product strategy, therefore no long term view of manufacturing's technological capabilities can be formed.
- There is genuine uncertainty as to how markets will change and how technology developments will affect product design.

All these problems do not diminish the need for setting manufacturing objectives but they do affect the nature of the task. It is rarely a mechanistic process where a few simple rules translate a statement of competitive strategy into a precise set of manufacturing objectives. Rather the process is one of exploration and compromise, best pursued in a workshop setting and involving all concerned parties. In such a setting data can be pooled, possible future scenarios debated and common ground defined. The end products however must be

- a clear ranked set of competitive performance objectives for each product or product group; and

- a view of the future which distinguishes between those capabilities the manufacturing function will *definitely* have to develop, those it definitely will *not* have to develop, and those which it *might* need to develop.

Objectives should be determined by customer needs

Marketing, Manufacturing and Product Development might be the protagonists physically present during this process, but the most important group – there only in spirit – are the organisation's customers. Chapter 1 raised this point. Customers define totally and absolutely what is important for Manufacturing. Their needs must be translated, through the manufacturing strategy, directly to the shop floor. Customer's priorities should be Manufacturing's priorities, customer's concerns, Manufacturing's concerns. In fact the success of this stage in strategy formulation is determined largely by its success in establishing a firm image of the customers, their needs, and what is required from manufacturing to satisfy them.

This boils down to assessing the relative importance of each of manufacturing's performance objectives. A process which is helped by using some kind of customer related importance scale on which the relative importance of each performance objective can be indicated for each product or product group. This involves revisiting the particularly useful idea of Order-Winning and Qualifying Objectives described in the first chapter.[1] As a reminder, customers can view each performance objective as,

Order winning objectives – those which, for the product–market combination being considered, directly influence the level of orders. They are the main performance indicators used by customers in the purchase decision. Or,

Qualifying objectives – those for which performance needs to be above a particular level for customers to even consider ordering. If a company's performance is below the 'qualifying' level it is unlikely to win orders. Once clearly above the 'qualifying' level, the marginal benefit of improvement to the company's competitive position is low. Or,

Less important objectives – which, as their name implies, are relatively unimportant compared to other performance objectives. Rarely do customers even consider these when making purchase decisions.

Figure 10.1 shows the general relationship between performance and competitive benefit for each of these three categories.

Within these categories though, there will be shades of importance. Both price and dependability, for example, may be order winning objectives in a particular market but they may not be of equal importance. More discrimination is needed in the scale.

The following is a nine point scale which can give a reasonable indication of the importance of each performance objective without excessive complication.

A 9-POINT IMPORTANCE SCALE
For this product/product group does this performance objective . . .

Order Winning Objectives

1. Provide a crucial advantage with customers – it is the main thrust of competitiveness.
2. Provide an important advantage with most customers – it is always considered.
3. Provide a useful advantage with most customers – it is usually considered.

Qualifying Objectives

4. Need to be at least up to good industry standard.
5. Need to be around the median industry standard.
6. Need to be within close range of the rest of the industry.

Less Important Objectives

7. Not usually come into customers' consideration, but could become more important in the future.

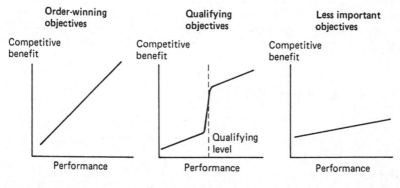

Figure 10.1: Manufacturing objectives can be classed as order-winning, qualifying, or less important

189

8. Very rarely come into customers' considerations.
9. Never come into consideration by customers or is ever likely to do so.

Example A small specialist manufacturer of industrial seals and sealing systems has patent protection for a particular seal which gives excellent long-life performance in corrosive environments. The design is radical and viewed with some suspicion by parts of the chemical industry to whom the product is sold. Nevertheless it has clear performance advantages over conventional seals, especially in technically difficult applications and commands a considerable price premium. Each product is designed for a specific customer application after an initial technical specification and quote has been prepared by the company's application engineers. Customers' delivery requirements can change at relatively short notice so as to fit into their construction or maintenance schedules.

After some debate the team who were responsible for this product identified manufacturing's performance objectives together with their importance to typical customers now and in the future (a three- to five-year horizon was used). This is shown in Figure 10.2, using the nine-point scale described above.

Of prime importance to most customers were the quality and versatility of applications engineering, though product quality (in terms of reliability) and a fast turn-round of initial enquiries were also considered important. Price was not a major competitive factor because of the product's considerable technical superiority, nevertheless it could become slightly more important as the performance of rival products improved. Similarly volume flexibility and delivery lead time, although not of great competitive significance at the moment, were considered likely to become more important. The arrows on Figure 10.2 indicate how the importance of each performance objective is likely to change over the period.

Step 2 – Judging Achieved Performance

If the customers are the silent presence during the definition of manufacturing objectives, competitors play the same role when assessing Manufacturing's achieved performance. Again, from Chapter 1, competitors provide a standard against which any manufacturing company should measure itself. In a strategic context performance measures only become meaningful when

190

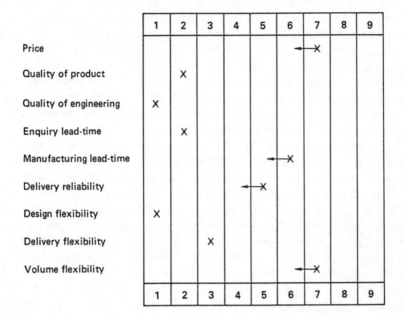

	1	2	3	4	5	6	7	8	9
Price						←X			
Quality of product		X							
Quality of engineering	X								
Enquiry lead-time		X							
Manufacturing lead-time						←X			
Delivery reliability					←X				
Design flexibility	X								
Delivery flexibility			X						
Volume flexibility						←X			
	1	2	3	4	5	6	7	8	9

Figure 10.2: The importance of each performance objective for the anti-corrosive seal

compared with competitors' achievements. And again each performance objective needs to be placed relative to competitors on some kind of scale. At its simplest the scale should reflect whether performance is better, the same, or worse than major competitors for each performance objective. Although again some more discrimination is often useful, such as in the following nine-point scale.

A 9 POINT PERFORMANCE SCALE
In this market sector, or for this product group, is our achieved performance in each of the performance objectives . . .

1. Consistently considerably better than our nearest competitor.
2. Consistently clearly better than our nearest competitor.
3. Consistently marginally better than our nearest competitor.
4. Often marginally better than most competitors.
5. About the same as most competitors.
6. Often within striking distance of the main competitors.
7. Usually marginally worse than main competitors.
8. Usually worse than most competitors.
9. Consistently worse than most competitors.

191

Example The manufacturer of industrial seals considered its performance relative to its competitors. The anti-corrosive seal had no direct competitors as such, but customers did have the option of using conventional seals and replacing them more frequently with more disruption to their process. This option was taken as the 'competitive position' against which the company measured its performance in terms of its performance objectives. Figure 10.3 shows the results of their deliberations.

The company was clear that its main superiority came from the strength of its applications engineering. It could consistently come up with innovative, high quality designs for a wide variety of applications. The major weakness of its design section was the length of time it took to respond to customers' enquiries. On the manufacturing side, finished quality was regarded as good, and some excess capacity in the machining cells gave the operation enough volume (range) flexibility to increase output substantially. Long supplier lead-times and subcontracting inflexibilities, however, made it difficult to change delivery dates once work had been scheduled. Even so, delivery reliability was poor compared to competitors. Not surprisingly, given its complexity, the product manufacturing cost was far higher than its less sophisticated competitors.

Judging performance means knowing competitors

The example of performance analysis shown in Figure 10.3 relies heavily on estimates of competitors' performance. This reflects the difficulty in getting hold of reliable information on the details of competitors' performance, especially such things as costs (because of an obvious desire for confidentiality), flexibility (because competitors probably don't know their own flexibility) and delivery dependability (because of both previous reasons). A rough idea of competitors' dependability can be estimated though by listening to customers and generally keeping one's ear to the ground. Delivery lead times and quality levels are, if anything, easier to judge.

Step 3 – Prioritising through the Importance/Performance Gap

It is the gap between the *importance* rating of each performance objective and its *performance* rating which gives the real clue to the priority it should be given. Neither a performance objective's importance nor its performance rating alone will do. For example,

	1	2	3	4	5	6	7	8	9
Cost*								X	
Quality of product			X						
Quality of engineering	X								
Enquiry lead-time							X		
Manufacturing lead-time							X		
Delivery reliability*						X			
Design flexibility*	X								
Delivery flexibility*								X	
Volume flexibility*			X						
*Estimated	1	2	3	4	5	6	7	8	9

Figure 10.3: The performance, relative to main competitors, of each performance objective for the anti-corrosive seal

delivery lead-time may be particularly important to customers, yet it will only deserve priority when improvement plans are being drawn up if performance is worse than competitors. Conversely, mix flexibility (which gives the ability to produce a wide product range) may be far worse than competitors, but why give it priority if customers are never likely to find product range important? Only by putting the two scales together can true priorities be judged. This is best done on the Importance/Performance matrix.

Figure 10.4 shows the Importance/Performance matrix. It takes the two scales developed in stages one and two. The 'importance' scale indicates how customers see the relative importance of each performance objective, the 'performance' scale rates each competitive performance objective against the levels achieved by competitors.

Remember, though, that neither scale is static, both rate positions relative to a dynamic external standard. Customers' preferences will change as markets develop and the economic environment changes. Competitors likewise are unlikely to stand still. They too will be striving to improve their performance. Any operation must be improving its own performance in absolute terms at least as fast as its competitors just to maintain its position on the performance scale. Improvement therefore doesn't just mean doing better than before, it means improving at a faster rate than competitors.

The Importance/Performance Matrix shown in Figure 10.4 is divided into four zones.

The 'Appropriate' zone This zone is bounded on its lower edge by a 'minimum performance boundary', that is, the level of performance below which the company, in the medium term, would not wish the operation to fall. Getting performance up to or above this boundary should be the first stage objective for any improvement programme. Performance objectives which fall in this area should be considered satisfactory, at least in the short to medium term. In the long term, however, most competitors will wish to edge performance towards the upper boundary of the zone. That, after all, is the ultimate aim – to be clearly better at everything.

The 'Improve' zone Any performance objective which lies below the lower bound of the 'appropriate' zone will be a candidate for

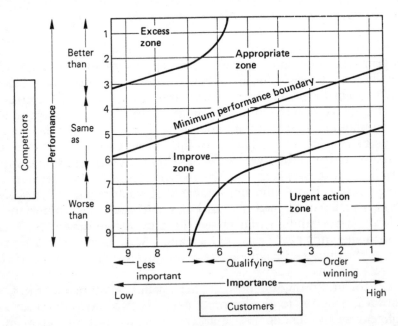

Figure 10.4: The importance/performance matrix compares what customers find important in achieved performance when compared with competitors

improvement. Those lying either just below the bound or in the bottom left corner of the matrix (where performance is poor but it matters less) are likely be viewed as non-urgent cases. Certainly they need improving, but probably not as a first priority.

The 'Urgent Action' zone More critical will be any performance objective which lies in the 'urgent action' zone. These are aspects of performance where achievement is so far below what it ought to be, given its importance to the customer, that business is probably being lost directly as a result. Short-term objectives therefore must be to raise the performance of any performance objectives lying in this zone at least up to the 'improve' zone, while in the medium term they need to be worked up beyond the lower bound of the 'appropriate' zone.

The 'Excess?' zone The question mark is important. If any performance objectives lie in this area their achieved performance is far better than would seem to be warranted. This does not necessarily mean that too many resources are being used to achieve such a level, but it may do. It is only sensible then to check to see if any resources used to achieve such a performance can be diverted into a more needy area; anything which falls in the 'urgent action' area, for example.

Example The corrosion resistant seal manufacturer could now position its operation on the Importance/Performance Matrix. This allowed the manufacturing team to debate the priorities for improvement implied by the positions of each performance objective, and set targets for short to medium term improvement. The arrows on Figure 10.5 show how the team set its improvement priorities.

First priority was given to improving enquiry lead-time and delivery flexibility, both of which had performance levels far enough below their importance ratings to put them in the 'urgent action' zone. Second priority was given to manufacturing lead-time and delivery reliability, both of which were forecast to increase in importance and were currently below industry average performance. Finally, manufacturing cost improvement was third priority. Not because performance was satisfactory (costs were well above the estimated costs of rival but inferior products, and prices even further above rival prices). Nor because cost was regarded as unimportant by the company (cost performance is never unimportant, even if it is not a major order winning factor). Rather the team argued that the superior technical performance of its product combined with a marketing concentration on those parts of the

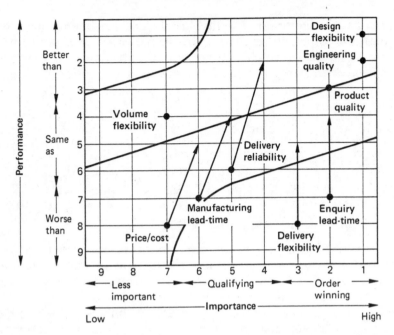

Figure 10.5: The importance/performance matrix for the anti-corrosion seal showing the improvement plans for the under-performing performance objectives

market where price was relatively unimportant, made high margins sustainable in the short term. Longer-term, however, price might become more important. Though even if it did not, the company would benefit from the reserves of competitiveness and higher margin which lower manufacturing costs would provide. Design flexibility, engineering quality and, to a lesser extent, product quality, were seen as giving the company strong competitive advantage. Nevertheless, long-term plans were established to maintain and improve their position.

Remember that this procedure is not intended to be entirely mechanistic. It does not totally prescribe priorities. It can, however, serve as a useful framework to raise the level of debate on current and future operations performance and relative priorities.

Moving customer perception is also an option

Shifting the position of performance objectives up the vertical scale on the importance/performance matrix is rightly seen as the major concern of the operations function. Short of unusually cooperative competitors, no one else can. Horizontal movement, however, is far from impossible, and can be a useful tactic. Horizontal movement on the matrix comes from changing customers' perceptions of what is important. Moving any performance objectives in the excess zone to the right in effect exploits a wasted capability by persuading customers to value it more highly than they would otherwise have done. Conversely, moving any performance objectives which lie in the urgent action area towards the left means that customers are being persuaded to regard some aspect of performance at which the operation does not excel as less important.

The major concern of course is always going to be to move performance vertically upwards. But companies who fail to at least explore the marketing dimension of competitive positioning, illustrated by horizontal movement on the matrix, are unnecessarily limiting their options.

Step 4 – Develop Action Plans

Charting the gap between performance objectives is an important guide to which, and how urgently, the various aspects of performance need to be improved. But it gives no indication of *how* performance is to be improved. Nor can any procedure be fully prescriptive at this stage. Every operation has its own characteristics and opportunities. Likewise every operation's personnel have the responsibility to generate an imaginative but practical set of action plans.

A useful approach is to examine the influence each of the activity areas has on each performance objective. Starting with the highest priority performance objectives this means asking what contribution to improvement could be derived from changes in,

- the operation's process technology;
- the operation's organisation and the development of its human resources;
- the operation's network of supply both in terms of information flow and material flow.

Example As part of its generation of action plans, the anti-corrosion seal team considered how it could improve its enquiry lead-time performance. A summary of the team's first thoughts emerged as follows.

Objective: Reduce enquiry lead-time to better than the industry average for standard seals.

Improvement Through Process Technology (in this case the Computer Aided Design (CAD) system)

- Enhance CAD system for 3D modelling
- Provide more work stations

Improvement Through Organisation and Human Resource Development

- Train more applications engineers for anti-corrosion seal work.
- Total reshaping of customer liaison procedures to allocate each job to a single engineer responsible for its progress.
- Redesign internal control procedures to (a) project manage all jobs, and (b) give total design approval to the engineer responsible.

Improvement Through the Supply Network

- Require the design engineer to quote an enquiry delivery date for every job.
- Develop a job tracking system showing the progress of every job in the enquiry stage.

Action plans need evaluating

Not all action plans may be worth pursuing. They need to be screened or evaluated first. The simplest way to do this is to use the three headings recommended in Chapter 7 to evaluate technology. As a reminder they are,

- the *Acceptability* of the plan. Does it do what it is supposed to do? Is it effective? How does it contribute to achieving the objective? Will the investment give a reasonable return? Does the plan conflict or support other proposed plans?
- The *Feasibility* of the plan. Is the financial investment beyond the budget limit? Does the operation have sufficient people with the appropriate skills to implement the plan? No matter how acceptable the idea, if it can't be implemented then it will provide nothing to overall performance improvement.
- The *Vulnerability* of the plan. What risks are inherent in the plan? What could go wrong? What is the downside risk of the

plan? What are the chances of things not going to plan and what would the consequences be?

The Implementation Agenda

Too often it all fouls up at the implementation stage. Strategy sets the destination but implementation defines how you get there – a more difficult task. Start by addressing the implementation agenda – the list of general questions, whose answers set the basic plan for implementation. The questions are

When to start?
Where to start?
How fast to go?
How to coordinate the improvement programme?

When to start There is only one absolute rule here. Don't start until all the issues on the implementation agenda have been addressed. Starting off without a reasonably clear idea of the migration path which bridges the performance gap is a sure way to reduce the chances of success. The implementation agenda in effect charts how progress is supposed to be made.

Some start times are nevertheless more propitious than others. Start when you can be sure of resourcing the improvement effort adequately with engineers and managers who really can devote enough time to the project. Don't start when capacity is seriously overstretched – all efforts will be on short-term output, not the programme. Start only after the last big push or project has settled down. Don't start when any major structural change (a new factory, product launch or process introduction for example) is imminent.

It would be foolish, though, to wait for all these conditions to be met. Any company which waits for ideal conditions is likely to be waiting a long time.

Where to start Two schools of thought here.

School One – start where you will get most direct benefit. Be rational and put effort and investment resources where the likely return, in terms of improved performance, will be maximised. This could be either where performance is so poor that relatively small improvement gives a disproportionate benefit, or it could be where the potential for dramatic improvement is high.

School Two – start where you believe there is the best chance of succeeding, preferably in a relatively calm backwater of the operation where any problems will not set back the whole programme. Start small and build up experience. Learn as you go, keep mistakes small scale and, above all, maintain credibility in the organisation. By the time the really important parts of the operation are being tackled, the programme team will have both the authority and the experience, if not to guarantee success, to make it far more likely.

Both approaches are perfectly legitimate, though the second has much to commend it when the improvement programme is inherently risky or when the implementation team have little experience of the type of change being made.

How fast to go? Managing the speed of improvement means understanding and nurturing the two modes of improvement: 'breakthrough improvement' and 'continuous improvement'.

Traditionally, the route to more effective manufacturing systems was seen as a series of 'breakthrough' step improvements, made as a result of correspondingly major changes to the system. Each change has a ratchet-like effect on the improvement process. Installing a new piece of plant, buying more sophisticated manufacturing control software, restructuring the shop floor organisation all could provide a significant turn to the improvement ratchet. After each improvement hike there should (preferably) come a time of calm and stability when the change can be bedded-in and any problems sorted out. In time the next step change will make a further turn in the improvement ratchet.

Seeing improvement as a series of sprints followed by recovery periods is certainly not the only way to achieve long-run success. Manufacturing systems also respond to a more gradual, incremental, but continuous, approach to improvement. Very few manufacturing systems do not have any potential for improvement within their existing structure. Tapping this potential improvement can have a long-term cumulative effect far greater than some more dramatic breakthroughs. This is the philosophy of continuous improvement introduced in Chapter 2. It shifts emphasis from the rate of improvement towards the momentum of the improvement process. It also changes the definition of success in the improvement process, from 'making big improvements' to 'always making some improvement' and failure from 'making small improvements', to 'failing to make some kind of improvement'.

These two approaches are sometimes presented as mutually exclusive. They are not. They just emphasise different sources of improvement potential. Both need exploiting to obtain maximum benefit. The continuous improvement philosophy needs to be embedded into the manufacturing consciousness so that every opportunity for improvement is taken. But tinkering with the system becomes progressively more difficult. Well planned radical step changes can open up opportunities for further continuous improvement as well as improving things in their own right. Moving manufacturing systems along the improvement path needs both tinkerers and dreamers.

How to co-ordinate the programme?

An improvement programme is like any other project – it needs managing.

First, the organisational environment in which the programme will need to fight for attention and resources should be predicted. With what conditions will the programme have to cope during its implementation? For example, are there product modifications or new product launches planned for the period? If so how will they interfere with the programme. Are there organisational or system changes planned by any other functions which could affect the programme? A little time spent planning at the start can prevent the kind of organisational surprises which can knock the programme off course.

Second, plan the resources which will be needed to staff the programme. Make sure that everyone with a part to play knows exactly what is required from them and approximately when this is likely to be.

Third, set milestones of progress in each area of the programme, so that feasibility can be judged at the planning stage and progress can be judged throughout the life of the programme.

Successful Implementation – The Key Elements

In the several studies of manufacturing implementation, both successful and unsuccessful, a number of key elements recur regularly, either as important prerequisites of success or as omissions which seriously harmed implementation success.[2]

Top management support – this always comes out as being important, especially at times of 'breakthrough' improvement where the 'champion' role requires top management to allocate and coordinate resources. Continuous improvement requires a different kind of support, emphasising a long-term continuing expectation of improvement.

Business driven – all manufacturing improvement is only a means to an end: improved competitiveness. The company's overall competitive imperatives must be clearly linked to every part of the improvement programme throughout its life.

Improvement drives technology – not the other way round. Competitiveness should drive improvement, so performance improvement determines the way technology is developed. Only firm-up on technology when the basic strategy and methodology of manufacturing are settled.

Back to basics rethinks – breakthrough improvements attempt significant step changes in performance and usually impact the whole manufacturing function. Don't miss the opportunity for a radical re-examination of the whole manufacturing system. Radical change isn't mandatory but a fundamental analysis of methods and objectives should accompany any step change implementation.

Change strategies are integrated – successful improvement programmes involve change over several fronts, technological, organisational, cultural. Only considering one aspect is too limiting a view. Integrating improvement strategies so that they support each other gets the operation 'firing on all cylinders'.

Invest in people as well as technology – organisations often seem strangely reluctant to invest in their human resources even a fraction of the cash they do in technology. Yet changes in methods, organisation or technology must be supported by changes in attitude by all employees – especially the managerial structure. Technological retooling must be accompanied by 'social retooling'.

Manage technology as well as people – conversely organisations often seem reluctant to 'manage' technology after the investment decision has been made. Technology needs integrating into the operation on strictly managerial criteria.

Everybody on board – any effective improvement must be understood and supported throughout the organisation, particularly the management structure. Without this, changes are implemented into 'unreformed' traditional structures, attitudes and work practices – where they often wither.

Clear explicit objectives – if you know what is expected of you it's easier to achieve it. Obvious maybe, but since manufacturing improvement usually involves cross-functional change the need for up-front communication of overall purpose becomes vital.

Time framed project management – keeping control is a prerequisite for maintaining support. Objective setting, schedules, resource plans, and milestones are as important here as for any other project.

PRACTICAL PRESCRIPTIONS

- The key test for any manufacturing strategy is that it connects the general activities of Manufacturing Operations to the overall competitive direction of the organisation. So always ask whether the manufacturing strategy gives useful guidance in making the many day-to-day and month-by-month decisions within the manufacturing function.
- Don't expect the process of putting manufacturing strategies together to be without its difficulties. Identify the likely difficulties and consider how they can be overcome. Frequently encountered problems are,

   Geographical dispersion of senior management;
   Who looks after the plant while strategy is being planned?;
   Manufacturing managers need help to think strategically.

- Check out any manufacturing strategy to see whether it is appropriate for its competitive objectives;

   comprehensive enough to provide guidance in all important areas;
   coherent enough to pull together the different parts of the strategy;
   consistent over time.

- Distinguish between the four distinct steps of strategy formulation – setting manufacturing objectives, judging current performance, identifying the priorities for improvement by referring to the gaps between objectives and performance, and developing and implementing action plans.
- The relative importance of each performance objective must be derived directly from customer needs.
- Different products or product groups should have different ranked sets of performance objectives if they compete in different ways.

- If some elements of competitive strategy are not clear then distinguish between those capabilities which the manufacturing function will definitely have to develop, those which it will definitely not have to develop and those it might have to develop. This at least clarifies the area of uncertainty.
- Classify each performance objective as either order winning, qualifying, or less important. Use some scale such as the nine point scale described previously to do this.
- Assess current achieved performance through comparison against competitors' performance.
- Initially judge performance as either better, about the same, or worse than competitors' for each performance objective. Then refine the judgment by using a scale such as the nine point performance scale described previously.
- Make sure that competitors' performance is tracked. Without knowledge of how they are performing the performance of one's own operation is unclear.
- Examine the gaps between what is important (reflecting customers' needs) and what performance is (reflecting competitors' performance) on an importance/performance matrix.
- Use the gaps between the importance and performance ratings of each performance objective to set its priority in the improvement plan.
- Always consider whether it is possible to exploit good performance and reduce the customer's view of poor performance by managing customer perceptions.
- Systematically look at how technology, the organisation and development of the operation's human resources, and its supply network, can contribute to improving performance.
- Set the implementation agenda by considering the following questions.
    When is the best time to start?
    Where in the organisation should improvement start?
    How fast should the programme develop?
    How is the whole improvement process to be coordinated?

NOTES AND REFERENCES

Chapter 1

1. Wickham Skinner launched many of the ideas which profoundly influenced the whole Manufacturing Strategy area. The best collection of his work is Skinner, W. *Manufacturing: the Formidable Competitive Weapon*, Wiley, 1985.

2. These two states of manufacturing performance, and those in between, are charted by Hayes, R.H., and Wheelwright, S.C. in *Restoring our Competitive Edge*, Wiley, 1984.

3. Terry Hill developed this idea and explains it in Hill, T., *Manufacturing Strategy*, Macmillan, 1985.

4. For more discussion see 'Competing with Tomorrow', *The Economist*, 12 May, 1990.

5. An idea fully developed in Prahalad, C.K. and Hamel, G., 'The Core Competence of the Corporation', *Harvard Business Review*, May–June, 1990.

6. Leadbeater, C., 'Piecing Together a Highly Complex Jigsaw Puzzle', Financial Times, 25 February 1991.

7. Nick Oliver likens the tone adapted by some proponents of fashionable (though not necessarily bad) ideas to evangelical religious meetings. His account of one management seminar holds many parallels with a revivalist meeting. Oliver, N., 'Just-in-time: the New Religion of Western Manufacturing', *British Academy of Management Conference Proceedings*, Glasgow, 1990.

Chapter 2

1. For a full description of this and other aspects of Hewlett-Packarrd's quality philosophy, see Rees, J. and Rigby P., 'Total Quality Control – the Hewlett-Packard Way', presented at the International Conference on TQM, London, IFS Publications, 1988.

2. The exact 'contents' of each category varies a little depending on which source is used. Take your pick from: British Standards Institution, 'BS6143', 1981; American Society for Quality Control, 'Quality Costs – What and How', 1981; Feigenbaum, A.V., *Total Quality Control*, McGraw-Hill, 1985.

3. Based on research by Dr Barrie Dale. For an excellent, if somewhat sceptical, review of the published evidence on quality-related costs, see Plunkett, J.J. and Dale, B.G., 'A Review of the Literature in Quality Related Costs', the *International Journal of Quality and Reliabiity Management*, Vol.4, No.1, 1987.

4. See Jones, A.K.V., 'Quality Management the Nissan Way' in *Managing Quality*, edited by B.G. Dale and J.J. Plunkett, Philip Allan, 1990, Also well worth reading for an insight into this company's ideas, Wickens, P., *The Road to Nissan*, Macmillan, 1987.

Chapter 3

1. Roy Merrill, of Northern Telecom Ltd, from 'How Northern Telecom competes on time', *Harvard Business Review*, July–August, 1989.

2. The concept of P:D ratios comes originally from Shingo, 'Study of Toyota Production Systems', Japan Management Association, 1981, and extended by Hal Mather in *Competitive Manufacturing*, Prentice Hall, 1988.

3. Figures taken from a revealing survey by Professor Colin New of Cranfield School of Management. New, C.C. and Myers, A., *Managing Manufacturing Operations in the UK 1975–1985*, the British Institute of Management, 1986.

4. From 'Britain's Six Best Factories'. *Management Today*, September 1988.

5. Schumann, M., 'Just-in-time manufacturing – effects on costs and logistics, Avon Cosmetics', proceedings of the third International Conference on Just-in-time Manufacturing, *IFS Publications*, 1988.

6. Information taken partly from Kehoe, L., 'The fruits of flexibility', *Financial Times*, 17th October 1990.

7. In case these figures seem unrepresentative, they are exactly the average lead-times in the two surveys discussed in New and Myers (op. cit.).

8. *The Economist*, 16 March 1991.

9. An example taken from Kim Clark and Takahiro Fujimoto, 'Overlapping Problem Solving in Product Development', in Ferdows, K., *Managing International Manufacturing*, Amsterdam, North

Holland, 1989. It is also discussed in Womack, J.P., Jones, D.T., and Roos, D., *The Machine that Changed the World*, Rawson Associates, 1990.

Chapter 4

1. New, C.C. and Sweeney, M.T., 'Delivery Performance and Throughput Efficiency in UK Manufacturing Industry'. *International Journal of Physical Distribution and Materials Management*, Vol.14, No.7, 1984.

2. Again, an example from New and Sweeney's excellent study (op. cit.).

3. Kasra Ferdows and Arnoud De Mayer at INSEAD in Fontainbleau have found evidence from their survey work that lasting improvement is built on Quality and Dependability. See 'Lasting Improvement in Manufacturing Performance: In search of a new theory', INSEAD Working Paper, 1990.

4. For a fuller description see, Withers, L., 'How to Prevent Downtime', *Industrial Computing*, October 1989.

Chapter 5

1. De Meyer, A. 'Flexibility, the next competitive battle', INSEAD Working Paper 86/31, Fontainebleau, 1986.

2. *The Economist*, 'Body building without tears', 21 April, 1990.

3. This example is described in more detail by John Dunn in 'Stiff Measures to Save Scotch', *The Engineer*, 29 March 1990.

4. For more on these aspects of flexibility see Slack, N., 'Focus on Flexibility', in Wild, R. (Ed.), *International Handbook of Production/Operations Management*, Cassell, 1989.

Chapter 6

1. Based on Eilan, S., and Gold, B., *Productivity Measurement*, Pergamon Press, 1978.

2. See for example Skinner W., 'The Productivity Paradox', *Harvard Business Review*, July–August, 1985.

3. This and other examples taken from 'Cost Cutting: How to do it Right', *Fortune*, April 1990.

4. In some industries which employ integrated processes, this phenomenon can be formalised. For example, parts of the chemical processing industry use the 'six tenths rule', which predicts that costs will rise by 60 per cent for every doubling of capacity.

5. Mather, H., *Competitive Manufacturing*, Prentice Hall, Englewood Cliffs, 1988, explains this with great insight.

6. New, C.C. and Myers, A., *Managing Manufacturing Operations in the UK*, BIM, 1986.

7. See for example Hayes, R., and Clark, K., 'Why Some Factories are More Productive Than Others', *Harvard Business Review*, Sept–Oct, 1986, or Schmenner, R., 'Comparative Factory Productivity', Duke University, July 1986.

8. This research is reported in Ferdows, K. and De Meyer, A., 'Lasting Improvement in Manufacturing Performance: In search of a new theory', INSEAD Working Paper, INSEAD Fontainebleau, 1989. The original model has been adapted slightly for the purpose of this book so as to use terminology used in other chapters and to reflect the author's description more precisely.

Chapter 7

1. From Schonberger, R.J., *World Class Manufacturing*, Free Press, 1986.

2. Dempsey, P., 'New corporation perspectives in FMS', *FMS Conference proceedings*, IFS, Kempstow, UK, 1983.

3. Bessant, J. and Haywood, B., 'The Introduction of Flexible Manufacturing Systems as an example of Computer Integrated Manufacturing', Innovation Research Group, Brighton Polytechnic, UK, 1985.

4. Tombak, M. and De Meyer, A., 'How the Managerial Attitudes of Firms with FMS Differ from Other Manufacturing Firms', INSEAD Working Paper No.86/15, 1986.

5. A fuller description of the use of robots in the Black and Decker plant at Spennymoor is given in, Mancey, J., 'Black and Decker's Workmate', *Industrial Computing*, April 1990.

6. A view eloquently described in *World Class Manufacturing* (op. cit.).

7. The idea is credited to two Harvard Business professors: Hayes, R., and Wheelwright, S., *Restoring our Competitive Edge*, Wiley, 1984.

8. Quote from the 'founder' of Manufacturing Strategy, Wickham Skinner, See *Manufacturing: the formidable competitive weapon*, (op. cit.).

9. This approach is fully explained in more general terms in Cooke, S. and Slack, N., *Making Management Decisions*, 2nd edition, Prentice Hall, 1991.

10. Kaplan, R.S., 'Must CIM be justified by faith alone?' *Harvard Business Review*, March–April 1986.

11. Another Kaplan idea (*see above*).

12. See Kim Clark and Takahiro Fujimoto, 'Overlapping Problem Solving in Product Development' in Ferdows, K. *Managing International Manufacturing*, Amsterdam, North Holland, 1989.

13. See Stinson, T., 'Team working in Real Engineering', *Machine Design*, March 1990.

Chapter 8

1. Dick Chase has taken these ideas further in his article 'The Service Factory', *Harvard Business Review*, 1989.

2. 'Britain's six best factories' *Management Today*, September 1988.

3. Voss, C., 'Managing New Manufacturing Technologies', Operations Management Association Monograph, No.1, 1986.

4. Parnaby, J., 'The Design of Competitive Manufacturing Systems', *International Journal of Technology Management*, Vol.1, Nos. 3/4, 1986.

5. See Bull, C., 'Integrating Maintenenace Production to Produce a Manufacturing Organisation', presented at the 13th Annual Maintenance Show, 1990.

6. Roberts, E.B., and Froham, A.C., 'Strategies for Improving Research Utilization', *Technology Review*, Vol. 80, No. 5, 1978.

7. 'Dow draws its matrix again – and again, and again . . .' *The Economist*, 5 August, 1989.

8. Again, see Hayes, R.H., Wheelwright, S.C., and Clark, K., (op. cit.) for more details.

Chapter 9

1. Based largely on Jones, C., 'Cross Boundary Supply Chain Management', *Professional Engineer*, Vol.3, No.5, 1990.

2. 'It's in the Mail', *The Economist*, 2 March 1991.

3. This example is taken from an Esprit II project. In it a methodology to analyse supply network problems was devised by a collaborative team from Warwick Business School and Lucas Industries. See Jones, C., and Clark, J., 'Effectiveness Framework for Supply Chain Management', *Journal of Computer Integrated Manufacturing Systems*, Vol.3, No.4, 1990.

4. The factors used here are based on those used by Professor Richard Lamming of Bath University (UK), but adapted slightly to fit in with the terminology used in other parts of the book. See Lamming, R., *Towards Best Practice*, Science Policy Research Unit, 1987.

5. See Lascelles, D.M. and Dale, R.G., 'Product Quality Through Supplier Development', in *Managing Quality*, edited by Dale, B.G. and Plunkett, J.T., Philip Allan, 1990.

6. Lamming, R., 'Strategic Trends in the Global Automotive Components Industry: The implications for Australia', Federation of Autoparts Manufacturers, Canberra, Australia, September 1990.

7. This example and the lessons from it are taken from Harrison, A., and Voss, C., 'Issues in setting up JIT supply', *International Journal of Operations and Production Management*, Vol.10, No.2, 1988.

8. A point made eloquently by Uday Karmarkar in his article 'Getting control of Just-in-time', *Harvard Business Review*, Sept–October, 1989.

9. From a report by David Petty and John Sharp on a Science Engineering Research Council ACME Project, 1988.

10. Voss, C. and Clutterbuck, D., *Just in Time – a Global Status Report*, IFS Publications, UK, 1989.

11. These two dimensions and Figure 9.4 are derived from (but are different to) those in Karmarkar, op.cit. and Voss, C., and Harrison, A., 'Strategies for implementing JIT', in Voss (ed). *Just-in-time Manufacture*, IFS, 1988.

Chapter 10

1. See Hill, T. *Manufacturing Strategy* Macmillan, London, 1984, for a full description of 'Order Winning' and 'Qualifying' objectives.

2. Based on Tranfield, D., and Smith, S., 'A Strategic Methodology for Implementing Technical Change in Manufacturing', British Academy of Management Conference, 1987.

INDEX

action plans for improvement, 197–8
Advanced Manufacturing Technology, 82, 123–7
 evaluation of, 132–6
Albion Pressed Metals, 95
Alco, lawn mowers, 95
Apple Computer, 53
appraisal costs, 28
automation, 120–3
automotive, supply chains, 165
Avon Cosmetics, 51

batching, 55
benchmarking, 6, 60
Birds Eye Walls, 153
Black and Decker, 122–3
BMW, 125–7
bottlenecks, 55
breakdowns, 69–72

capacity,
 increment of, 117–20
 overloading, 73
 utilisation, 98–9
changeover flexibility, 11–12, 94–5
Chevrolet, 139
Colgate Palmolive, 71
competitors' performance, estimating, 5
Computer Integrated Manufacturing, 17
concurrent engineering, 139

conformance to specification, 34, 36
continuous improvement, 31–2, 42, 200
core competencies, 6
cost advantage, the, 8, 97–116
 breaks, 103
costs, strategic determinants of, 101
cost-volume relationship, 102–3
Cummins Diesel Engines, 95
customers,
 contact with factory, 144–5
 influence on strategy formulation, 188
 moving perceptions of, 196–7
 role in manufacturing, 3
 voice of, 4

decision making, simplified, 55, 58
delivery
 flexibility, 83
 integrity, 64
Dell Computers, 162
Delta Crompton Cables, 13
demand
 chase strategies, 108
 variation, influence on cost, 107–9
dependability
 advantage, the, 8, 61–76
 definitions of, 63

211